Vibrations and stability of multiple parameter systems

Monographs and textbooks on mechanics of solids and fluids

editor-in-chief: G. Æ. Oravas

Mechanics of elastic stability

editor: H. Leipholz

1. H. Leipholz
   Theory of elasticity

2. L. Librescu
   Elastostatics and kinetics of anisotropic and heterogeneous shell-type structures

3. C. L. Dym
   Stability theory and its applications to structural mechanics

4. K. Huseyin
   Nonlinear theory of elastic stability

5. H. Leipholz
   Direct variational methods and eigenvalue problems in engineering

6. K. Huseyin
   Vibrations and stability of multiple parameter systems

Vibrations and stability of multiple parameter systems

K. Huseyin

University of Waterloo
Faculty of Engineering
Department of Systems Design
Waterloo, Ontario, Canada

NOORDHOFF INTERNATIONAL PUBLISHING
ALPHEN AAN DEN RIJN

ISBN 90 286 0136 8

Printed in the Netherlands

*"Stabilität oder Instabilität,
das ist die Frage..."*

Contents

Preface

Systems are usually under the influence of several independent parameters, and starting from an initially stable state it is often required to determine or predict the behaviour of a given system when these parameters are varied. If the system possesses a potential (gradient system), such as the total potential energy associated with conservative problems, the behaviour of the system can conveniently be studied via statical methods. In the case of non-gradient systems, however, the stability investigations often require dynamical methods. In fact, LYAPUNOV's definition of stability is a dynamical one, and it is only natural to employ dynamical criteria in most general situations. This book is mainly concerned with the dynamical behaviour and stability of autonomous systems, and it extends some of the basic concepts introduced in an earlier monograph (Nonlinear Theory of Elastic Stability) with regard to conservative systems, to non-gradient systems.

The main features of the present book may be summarized as follows:

1. It provides a general linear theory concerning the free vibrations and stability of discrete autonomous multiple-parameter systems.

2. The systems under consideration are further divided into certain well-defined classes, and various phenomena related to vibrations and stability of these classes of systems are exposed systematically. In this manner, several theorems concerning the basic properties and characteristic features of each class are formulated. These theorems have practical as well as theoretical value, and are often associated with lower and/or upper bound estimates of the characteristic surface and stability boundary.

3. Although examples are all drawn from mechanics, the general nature of formulation allows for application of the theory to problems in a variety of disciplines, provided the essential variables and parameters are interpreted appropriately. Indeed, this is part of our underlying motivation.

It is therefore, expected that the content of this book will be of interest to students and researchers in all branches of engineering as well as applied mathematics and physics in general. The book is suitable for adoption as a text at graduate or senior undergraduate levels.

The development of the theory relies heavily on matrix theory, and the first chapter is intended as a fairly self-contained concise reference on this subject. The content of the chapter is adjusted according to the requirements of the succeeding chapters, and a few relatively recent developments, which are not yet available in standard texts, are also included. The exercises at the end of appropriate sections are designed as an integral part of the treatment and are referred to freely as required. The reader, however, may start reading the book from the second chapter which is concerned with basic definitions and stability criteria. It also provides an appropriate classification of the systems to be considered. The remaining chapters are developed along the directions indicated in chapter 2, and deal with well-defined classes of systems, each chapter starting with a general discussion and ending with a brief section concerning the effect of dissipative forces on the stability behaviour. Examples are chosen specifically for illustration of various concepts, methods and phenomena. It is assumed that the reader has the usual background in calculus and mechanics expected from an engineering student at about the senior level.

I would like to express my gratitude to Dr. H. Leipholz, the editor of this series, for his encouragement. I am indebted to Dr. R. Plaut, who read the manuscript carefully and made numerous valuable suggestions. Thanks are also due to Dr. V. Mandadi who helped prepare the manuscript and Mrs. Liz Houghtling who typed the manuscript so skillfully. Finally, the support of the National Research Council of Canada and Alexander von Humboldt-Stiftung, FRG, is acknowledged.

Darmstadt
May 1977

Koncay Huseyin

1

Mathematical preliminaries

1.1 Vectors and matrices

In this book matrix theory will provide most of the essential tools of analysis, and the present chapter is intended as a concise and convenient reference for the development of the stability theory in the succeeding chapters. Clearly, in such a limited space a systematic treatment of matrix theory cannot be expected; indeed, only selected topics relevant to the content will be included. There are many excellent books on matrices and the reader is referred to the standard works by Gantmacher [1], Bellman [2] and Lancaster [3] among several others for a thorough treatment of the subject. The reader is assumed to be familiar with the subject including determinants as well as having a solid foundation in calculus.

A matrix is a rectangular array of numbers or functions represented by the symbol

$$A = \begin{bmatrix} a_{11} & a_{12} & \cdots & a_{1n} \\ a_{21} & a_{22} & \cdots & a_{2n} \\ \cdots & \cdots & \cdots & \cdots \\ a_{m1} & a_{m2} & \cdots & a_{mn} \end{bmatrix}.$$

The numbers or functions a_{ij} of this array are called the *elements* of A and can have real or complex values. The matrix A has m *rows* and n *columns*, and is said to have the *order* $m \times n$. When $m = n$ the matrix is *square* of order n.

The $(n \times m)$ matrix with elements a_{ji} obtained from the $(m \times n)$ matrix A with elements a_{ij}, by interchanging the rows and columns of A, is called the *transpose* of A and is denoted by A'.

Two matrices A and B are said to be equal, $A = B$, when they have the same order and $a_{ij} = b_{ij}$ for each pair of subscripts i and j. The following operations will now be defined:

(1) Addition: If both A and B have the same order, the sum $A + B$ of these matrices is defined as the matrix C of the same order and with

elements

$$c_{ij} = a_{ij} + b_{ij}.$$

It can readily be shown that the addition defined above is commutative and associative:

$$A+B=B+A,$$

$$(A+B)+C=A+(B+C).$$

Note that the subtraction of B from A can be defined as $A+(-)B$, and if $A=B$ this operation results in the *zero matrix* 0 all of whose elements are zero.

(2) Product of a matrix with a scalar: The product of A with a scalar α from the real or complex field F, αA, is another matrix with the elements αa_{ij}. This operation is commutative, associative and distributive both with respect to addition of matrices and scalars:

$$\alpha A = A\alpha,$$

$$\alpha(\beta A) = (\alpha\beta)A,$$

$$\alpha(A+B) = \alpha A + \alpha B,$$

$$(\alpha+\beta)A = \alpha A + \beta A.$$

Under the operations (1) and (2) the set of all $m \times n$ matrices with elements from the field F forms a *linear space* over the field F.

(3) Matrix product: The product of the $m \times p$ matrix A and the $p \times n$ matrix B is defined to be the $m \times n$ matrix $C=AB$ whose elements are given by

$$c_{ij} = a_{ik} b_{kj}$$

where the *summation convention* is adopted. Thus, while the indices i and j *identify* the element, the expression is summed over the repeated index k for all admissible values of the index. This convention will be used throughout the book.

It is observed that matrix multiplication is only meaningful for *conformable* matrices such that the number of columns of the first matrix is equal to the number of rows of the second. Furthermore, in general, this operation is not commutative but it is associative and distributive:

$$AB \neq BC,$$

$$(AB)C = A(BC),$$

$$(A+B)C = AC + AC.$$

When matrices A and B are such that $AB = BA$ they are said to *commute.* It follows immediately that if A and B commute they must be square matrices of the same order.

From the definitions of the operations (1), (2) and (3) and of the transpose of a matrix it can be shown that

$$(A+B)' = A' + B',$$
$$(\alpha A)' = \alpha A',$$
$$(AB)' = B'A',$$
$$(A')' = A.$$

A matrix of order $n \times 1$, consisting of a single column, is called a *column vector* of order n,

$$x = \begin{bmatrix} x_1 \\ x_2 \\ \cdot \\ \cdot \\ \cdot \\ x_n \end{bmatrix}.$$

Similarly, a matrix of order $1 \times n$, consisting of a single row, is called a *row vector* of order n,

$$x' = (x_1,\ x_2,\ \ldots,\ x_n).$$

Clearly, the transpose of a column (row) vector is a row (column) vector. In general, a vector can be defined as an ordered n-tuple of real or complex numbers, and the operations defined for matrices apply to column and row vectors as well. The matrix product, however, can now be achieved in two ways. Thus, the *inner product* (which is also known as the scalar or dot product) is defined as:

$$\langle x, y \rangle = x'y = (x_1,\ x_2, \ldots,\ x_n) \begin{bmatrix} y_1 \\ y_2 \\ \cdot \\ \cdot \\ \cdot \\ y_n \end{bmatrix} = x_1y_1 + x_2y_2 + \cdots + x_ny_n. \qquad (1.1)$$

On the other hand, the *outer product* is defined as:

$$xy' = \begin{bmatrix} x_1 \\ x_2 \\ \cdot \\ \cdot \\ \cdot \\ x_n \end{bmatrix} (y_1, y_2, \ldots, y_n) = \begin{bmatrix} x_1y_1 & x_1y_2 & \cdots & x_1y_1 \\ x_2y_1 & x_2y_2 & \cdots & x_2y_n \\ & \cdots\cdots & & \\ x_ny_1 & x_ny_2 & \cdots & x_ny_n \end{bmatrix} \tag{1.2}$$

It is noted that the inner product is a scalar (an element from the field F) while the outer product is a matrix of order n.

The following properties of the inner product

$$\langle x,y\rangle = x_iy_i \tag{1.3}$$

can be derived directly from its definition:

$$\left.\begin{aligned} &\langle x,y\rangle = \langle y,x\rangle \\ &\langle x,\alpha y+\beta z\rangle = \alpha\langle x,y\rangle + \beta\langle x,z\rangle. \end{aligned}\right\} \tag{1.4}$$

It is important to note that the inner product $\langle x, x\rangle$ can be considered to represent the square of the *length* of a *real* vector x, borrowing the illustrative language of geometry associated with the three dimensional real space R_3. Thus

$$\langle x, x\rangle^{\frac{1}{2}} = \|x\| \quad \text{where } \|x\| \text{ denotes the length of } x.$$

Evidently, for a real vector x,

$$\left.\begin{aligned} &\langle x, x\rangle \geq 0 \\ \text{and} \\ &\langle x, x\rangle = 0 \quad \text{if and only if} \quad x = 0. \end{aligned}\right\} \tag{1.5}$$

If the inner product of two *real* vectors x and y is zero,

$$\langle x, y\rangle = 0, \tag{1.6}$$

they are said to be *orthogonal.*

In order to extend the concept of orthogonality to complex vectors, first note that when x is a complex vector the inequality (1.5) may not hold, permitting a complex vector to be orthogonal to itself. This peculiarity may be avoided if the inner product of x with its complex conjugate $\bar{x}$ is considered, since

$$\langle \bar{x}, x\rangle = \bar{x}_1x_1 + \bar{x}_2x_2 + \cdots + \bar{x}_nx_n \geq 0$$

and equality holds if and only if $x = 0$. The proper concept of orthogonality for complex vectors, therefore, is expressed by the relation

$$\langle \bar{x}, y \rangle = 0. \tag{1.7}$$

Taking the complex conjugate transpose of this equation,

$$\langle \bar{x}, y \rangle = \langle \bar{y}, x \rangle = 0,$$

one observes that it does not matter which of two vectors is used first in testing for orthogonality*.

A vector x is called *normalized* (or is said to be a *unit vector*) if the *length*

$$\|x\| \equiv \langle \bar{x}, x \rangle^{\frac{1}{2}} = 1. \tag{1.10}$$

Any nonzero vector x can be associated with a unique unit vector by dividing each element of x by the length of the vector. This process is called *normalization.*

An $n \times n$ matrix is called a *square matrix.* If all the elements off the *main diagonal* (consisting of the elements a_{ii}) of a square matrix are zero it is called a *diagonal matrix* ($a_{ij} = 0$ for $i \neq j$). A *unit matrix* is a special diagonal matrix whose elements are ones, and is denoted by I. A diagonal matrix is usually expressed as diag $[a_{11} \quad a_{22} \cdots a_{nn}]$.

A square matrix A is said to be *singular* if its determinant, $|A|$, is zero. If $|A| \neq 0$, then, the matrix is *nonsingular.* The n-square matrix B is

* Many authors (e.g., [1], [3]) define the inner product of two vectors (real or complex) on the basis of the scalar used for the definition of orthogonality. Thus, the inner product of two vectors x and y is defined (note the order) by

$$\langle x, y \rangle = \bar{x}_1 y_1 + \bar{x}_2 y_2 + \cdots + \bar{x}_n y_n \tag{1.8}$$

in which case

$$\langle x, y \rangle = \overline{\langle y, x \rangle} \quad \text{(where bar indicates a complex conjugate)} \tag{1.9}$$

since

$$\langle x, y \rangle = \bar{x}_i y_i = y_i \bar{x}_i = \overline{\bar{y}_i x_i} = \overline{\langle y, x \rangle}.$$

Under this definition the scalar product $x'y = x_i y_i$ in general is not an inner product unless both x and y are real, and the notation $\langle\rangle$ is not used. This definition, however, will not be adhered to in this book and inner product will not be distinguished from the scalar product.

said to be the *inverse* of the n-square matrix A if $AB = BA = I$; B is denoted by A^{-1}. A^{-1} exists only if A is nonsingular. For, from $AA^{-1} = I$ one has $|A|\,|A^{-1}| = 1$ which cannot hold if $|A| = 0$. It can further be shown that

$$A^{-1} = \frac{\text{adj } A}{|A|} \tag{1.11}$$

where adj A denotes the *adjoint* of A with the elements α_{ji}. The definition of adj A is based on the cofactors of the elements a_{ij} of the matrix A. The determinant $|m_{ij}|$ of the $(n-1)$-square matrix m_{ij}, obtained by deleting the ith row and jth column of A, is called the *minor* of a_{ij}; the signed minor, $(-1)^{i+j}\,|m_{ij}|$, is called the *cofactor* of a_{ij} and is denoted by α_{ij}.

Since

$$a_{ik}\alpha_{kj} = |A|\,\delta_{ij}, \tag{1.12}$$

equation (1.11) follows. Here δ_{ij} is the Kronecker delta with the following properties:

$$\begin{aligned} \delta_{ij} &= 1 \quad \text{if} \quad i = j \\ \delta_{ij} &= 0 \quad \text{if} \quad i \neq j. \end{aligned} \tag{1.13}$$

From the definitions, it can be shown that

$$(A^{-1})' = (A')^{-1}$$

and

$$(AB)^{-1} = B^{-1}A^{-1}.$$

Certain special matrices will now be defined. A square matrix A is called *symmetric* if $A = A'$. A square matrix A is called *skew-symmetric* if $A = -A'$. It follows that $a_{ij} = -a_{ji}$ for all values of i and j, and the diagonal elements of a skew-symmetric matrix are all zero.

A square matrix A is called *orthogonal* if $A'A = I$; i.e., $a_{ik}a_{kj} = \delta_{ij}$. In view of the definition of A^{-1}, one observes that $A' = A^{-1}$.

The complex conjugate of a complex matrix A (with complex elements a_{ij}) is defined as a matrix whose elements are $\bar{a}_{ij}$ and is denoted by $\bar{A}$.

A square matrix A is called *Hermitian* if $\bar{A}' = A$. The diagonal elements of an Hermitian matrix are real since $\bar{a}_{ii} = a_{ii}$.

A square matrix A is called *skew-Hermitian* if $\bar{A}' = -A$. Clearly, the diagonal elements of a skew-Hermitian matrix are either zeroes or pure imaginaries.

It is noted that a real symmetric matrix is a special kind of Hermitian matrix; in fact, a real Hermitian matrix is symmetric. Also, all the important properties of symmetric matrices can be extended to Hermitian matrices.

A square matrix A is called *unitary* if $\bar{A}'A = I$.

It is noted that every square matrix A can be expressed as the sum of a symmetric matrix $B = \frac{1}{2}(A + A')$ and a skew-symmetric matrix $C = \frac{1}{2}(A - A')$. Similarly, every square matrix with complex elements can be written as the sum of an Hermitian and skew-Hermitian matrix.

The sum of the elements on the main diagonal of a square matrix A is called the *trace* of A,

$$\operatorname{tr}(A) = a_{11} + a_{22} + \cdots + a_{nn}.$$

An *upper* (*lower*) *triangular* matrix is one which has nonzero elements only on or above (below) the main diagonal; i.e., if $a_{ij} = 0$ for $i > j$ ($a_{ij} = 0$ for $j > i$) the square matrix is upper (lower) triangular.

Finally, the important concepts of linear dependence of vectors and the rank of a matrix will be defined.

The vectors $x^1, x^2, \ldots, x^n$ over the field F (complex or real) are said to be *linearly dependent* if there exist n elements $\alpha_1, \alpha_2, \ldots, \alpha_n$ of F, not all zero, such that

$$\alpha_1 x^1 + \alpha_2 x^2 + \cdots + \alpha_n x^n = 0. \tag{1.14}$$

Otherwise the n vectors are *linearly independent.* If the n vectors are linearly dependent, some one of them can be expressed as a linear combination of others.

An $m \times n$ matrix A is said to have *rank* r if the order of its largest non-vanishing minor is r. In analogy with the minor (of order one) defined before, the minor of order k of an $m \times n$ matrix A is the determinant of the k-square matrix obtained by deleting any $m - k$ rows and $n - k$ columns of A. It can be shown that if the largest number of linearly independent column (or row) vectors of A is r, then the rank of A is also r.

An n-square matrix A is obviously nonsingular if its rank is n. If the rank $r = n - \gamma$, where γ is positive ($\gamma < n$), the matrix A is said to have *nullity* (or *degeneracy*) γ. In this case, there are γ independent linear relations between the columns of A. For a system of homogeneous equations $Ax = 0$, the solution vectors x constitute the *null space* of A which has the dimension γ; i.e., if A has nullity γ, then $Ax = 0$ has γ linearly independent solutions $x^1, x^2, \ldots, x^\gamma$ such that every solution of the system $Ax = 0$ can be expressed as a linear combination of the x^i ($i = 1, 2, \ldots, \gamma$) and any linear combination is a solution. Obviously, if A

is nonsingular, then the only solution of $Ax=0$ is $x=0$. Also, note that a *vector space* E_n (real or complex) contains at most n linearly independent vectors, and any set of $n+1$ vectors is linearly dependent. A vector space is said to be *spanned* or *generated* by the n-vectors y^i $(i=1,2,\ldots,m)$ if every vector in that space can be expressed as a linear combination of the y^i.

The *Gram-Schmidt orthogonalization process* can be used to obtain an orthogonal set from n linearly independent vectors. If x^i $(i=1,2,\ldots,n)$ are real linearly independent vectors, define

$$y^1 = x^1,$$

$$y^2 = x^2 - \frac{\langle y^1, x^2\rangle}{\langle y^1, y^1\rangle} y^1,$$

$$y^3 = x^3 - \frac{\langle y^2, x^3\rangle}{\langle y^2, y^2\rangle} y^2 - \frac{\langle y^1, x^3\rangle}{\langle y^1, y^1\rangle} y^1,$$

etc.

The vectors y^i $(i=1, 2, \ldots, n)$ are evidently mutually orthogonal. An *orthonormal* set can immediately be obtained as

$$e^i = \frac{y^i}{\langle y^i, y^i\rangle^{\frac{1}{2}}} \qquad i=1, 2, \ldots, n$$

so that $\langle e^i, e^i\rangle = \delta_{ij}$.

Exercises

1. A rectangular matrix A can be partitioned into matrices of smaller order,

$$A = \begin{bmatrix} A_{11} & A_{12} \\ A_{21} & A_{22} \end{bmatrix},$$

for convenience. Show that the matrix product of two partitioned matrices $[A_{ij}]$ and $[B_{ij}]$ can be determined by

$$[A\,B]_{ij} = A_{ik}B_{kj},$$

assuming conformability as required.

2. $AA=A^2$, $A^2A=A^3$, etc., may be used without ambiguity. What can be said about $(T^{-1}AT)^n$? A^mA^n? $(A^m)^n$?

3. Define the matrix $e^A = I + A + \frac{1}{2!}A^2 + \cdots$ and discuss e^{At}, $e^{(A+B)t}$, and $e^{A+\varepsilon B}$ where t and ϵ are scalars and ϵ is small.

4. It is known from the theory of determinants that if the elements of an $n \times n$ determinant are differentiable functions of a variable λ, then the derivative of the determinant with respect to λ is the sum of n determinants obtained by replacing each column (or row) by its derivative with respect to λ. On the other hand, the derivative of the corresponding matrix is another matrix whose elements are the derivatives of the elements of A.

With these definitions, show that

$$\frac{\mathrm{d}^n}{\mathrm{d}t_n}\left[e^{tA}\right] = A^n \mathrm{e}^{tA} = \mathrm{e}^{tA} A^n$$

1.2 Bilinear and quadratic forms

An expression of the form

$$f = a_{ij} x_i y_j; \qquad i = 1, 2, \ldots, m, \qquad j = 1, 2, \ldots, m \tag{1.15}$$

is called a *bilinear form* in the two sets of variables x_i and y_j.

The scalar f is, in fact, the inner product of the vectors x and Ay (both of order m) and can be expressed as

$$f = \langle x, Ay \rangle.$$

Since

$$\langle x, Ay \rangle = x'Ay = y'A'x = \langle y, A'x \rangle, \tag{1.16}$$

the effect of A on y may be considered to be equivalent to the effect of A' on x as far as the inner product is concerned. A' can, therefore, be regarded as an *induced operator* or *adjoint operator.*

If A is symmetric so that $A = A'$, then

$$\langle x, Ay \rangle = \langle y, Ax \rangle$$

and A is *self-adjoint.*

A homogeneous polynomial of the type

$$f = a_{ij} x_i x_j; \qquad i = 1, 2, \ldots, n, \qquad j = 1, 2, \ldots, n$$

is called a *quadratic form* in the set of variables x_i. It can be expressed as

$$f = \langle x, Ax \rangle \tag{1.17}$$

where A is a square matrix. If A is skew-symmetric, then $\langle x, Ax \rangle = 0$ since taking the transpose of (1.17) yields

$$\langle x, Ax \rangle = -\langle x, Ax \rangle.$$

Recalling the fact that every square matrix can be expressed as the sum of a symmetric and skew-symmetric matrix, one may always assume that the matrix of a quadratic form is symmetric.

A real symmetric matrix A is said to be *positive definite* if the quadratic form

$$\langle x, Ax\rangle > 0$$

for all nonzero real vectors x. Similarly, a real symmetric matrix A is called *positive semi-definite* (or *nonnegative definite*) if

$$\langle x, Ax\rangle \geq 0$$

for all real x, equality holding for some nonzero vectors. As an extension of these definitions one may introduce *negative definite*, *negative semi-definite* and *indefinite* real symmetric matrices according to whether the corresponding quadratic form for all non-zero real vectors is negative, negative or zero, and negative or positive, respectively. The quadratic form associated with a positive definite matrix is said to be positive definite, etc.

It can be shown that if A is a real symmetric and positive definite matrix, then

$$\langle \bar{x}, Ax\rangle > 0 \tag{1.18}$$

for all complex vectors x other than the zero vector. To show this, first note that $\langle \bar{x}, Ax\rangle$ is a real scalar since it is equal to its complex conjugate. Expressing the complex vector x as $x = y + iz$ where both y and z are real, one has

$$\begin{aligned}\langle \bar{x}, Ax\rangle &= \langle (y - iz), A(y + iz)\rangle \\ &= \langle y, Ay\rangle + \langle z, Az\rangle > 0\end{aligned}$$

since A was assumed to be positive definite. Similarly, it can readily be shown that if A is real skew-symmetric, then $\langle \bar{x}, Ax\rangle$ is imaginary for all complex vectors x other than the zero vector.

It is also noted that positive definite matrices are necessarily nonsingular, because if they were singular, then $Ax = 0$ for some $x \neq 0$ and $\langle x, Ax\rangle = 0$.

Exercises

1. If x and y are two real vectors, show that (Cauchy's inequality)

$$\langle x, y\rangle^2 \leq \langle x, x\rangle\langle y, y\rangle,$$

equality holding if and only if x and y are linearly dependent.

2. Show that for any two complex vectors x and y one has

$|\langle \bar{x}, y\rangle|^2 \leq \langle \bar{x}, x\rangle \langle \bar{y}, y\rangle.$

3. The form

$h = \langle \bar{x}, Hx\rangle$

where H is Hermitian is called an Hermitian form. How would you define the positive definiteness of an Hermitian matrix and form? Express the analogues of the other properties of real symmetric matrices for Hermitian matrices.

4. If T is a real nonsingular matrix, show that $T'T$ is symmetric and positive definite.

1.3 Canonical forms of matrices

It is often desirable to express a matrix in the simplest possible form, and the transformations leading to such reductions will now be discussed.

Two matrices A and B are said to be *equivalent* if there exist nonsingular matrices P and Q such that

$$B = PAQ.$$

The rank of equivalent matrices is the same since pre-multiplication by P and post-multiplication by Q do not alter the linear independence of rows and columns of A.

Congruence

A special case of equivalence is *congruence.* Two n-square matrices A and B are called *congruent* if there exists a nonsingular matrix Q such that

$$B = Q'AQ.$$

Congruence arises, for example, in the transformation of quadratic forms. Consider the quadratic form

$$f = \langle x, Ax\rangle;$$

if one changes the coordinates by means of the linear transformation $x = Qy$, then

$$\begin{aligned} f &= \langle Qy, AQy\rangle = \langle y, Q'AQy\rangle \\ &= \langle y, By\rangle \end{aligned}$$

where the matrices A and B are congruent.

Congruence transformations preserve the symmetry of matrices. Thus, if A is symmetric, so is B since $B = B'$. Similarly, every matrix $B = Q'AQ$ congruent to a skew-symmetric matrix A is also skew-symmetric.

Similarity

Two n-square matrices A and B are *similar* if there exists a nonsingular matrix T such that

$$B = T^{-1}AT.$$

This transformation does not preserve symmetry but has other important properties which will be discussed in the sequel.

Orthogonal transformations

If A is a real square matrix and there exists an orthogonal matrix $P(P' = P^{-1})$ such that

$$B = P^{-1}AP = P'AP,$$

then B and A are *orthogonally* similar. Actually, the orthogonal transformation combines congruence and similarity and their properties.

Orthogonal transformations leave distances invariant. The length of a vector x, for example, remains the same after the orthogonal transformation $y = Px$, since

$$\langle y, y\rangle = \langle Px, Px\rangle = \langle x, P'Px\rangle = \langle x, x\rangle.$$

Eigenvalues and eigenvectors

Eigenvalues and eigenvectors play a significant role in transforming a matrix into a simpler form, and they will be discussed here in this context.

If A is an n-square matrix, and q an n-vector, interest often focusses on those vectors which are transformed into scalar multiples of themselves by the matrix A such that

$$Aq = \lambda q, \tag{1.19}$$

where λ is a scalar, or equivalently

$$(A - \lambda I)q = 0. \tag{1.20}$$

Nontrivial solutions of this equation exist only if the matrix $A - \lambda I$ is singular. In other words, λ must be a root of the determinantal equation

$$|A - \lambda I| = 0. \tag{1.21}$$

This equation is called the *characteristic equation* of the matrix A. The polynomial in λ, resulting from the expansion of the determinant (1.21), is known as the *characteristic polynomial* of A. The roots of this polynomial are the *eigenvalues* or *characteristic roots* of A, which can be *distinct* or *coincident* (simple or multiple).

The nontrivial solution of (1.20) corresponding to a given eigenvalue is a *right eigenvector* of A. Similarly,

$$r'(A - \lambda I) = 0$$

yields a *left eigenvector* for the same eigenvalue. Clearly, the left and right eigenvectors of a symmetric matrix are the same. Associated with each *distinct* eigenvalue, there is a right (and left) eigenvector. If λ is a multiple root of an arbitrary square matrix, however, there may not be an equal number of distinct eigenvectors. On the other hand, real symmetric matrices of order n have n distinct eigenvectors, as will be shown later.

Real symmetric matrices

There are two fundamental properties of real symmetric matrices:

1. The eigenvalues of a real symmetric matrix are real.

To prove this, let A be a real symmetric matrix with the eigenvalue λ. Then, pre-multiplying the equation

$$Aq = \lambda q$$

by $\bar{q}'$, one has

$$\langle \bar{q}, Aq \rangle = \lambda \langle \bar{q}, q \rangle.$$

But both $\langle \bar{q}, Aq \rangle$ and $\langle \bar{q}, q \rangle$ are real and hence λ is real.

2. Eigenvectors associated with distinct eigenvalues of a real symmetric matrix are orthogonal.

It is first noted that, since the eigenvalues are real, eigenvectors are also real. To prove orthogonality, let λ_i and λ_j be two distinct eigenvalues of the real symmetric matrix A. Then the equations

$$Aq^i = \lambda_i q^i,$$
$$Aq^j = \lambda_j q^j, \qquad \lambda_i \neq \lambda_j$$

lead to

$$\langle q^j, Aq^i \rangle = \lambda_i \langle q^j, q^i \rangle$$
$$\langle q^i, Aq^j \rangle = \lambda \langle q^i, q^j \rangle$$

which, upon subtraction, yield

$$0=(\lambda_i-\lambda_j)\langle q^i,\ q^j\rangle$$

since $\langle q^j,\ Aq^i\rangle=\langle q^i,\ Aq^j\rangle$. Therefore,

$$\langle q^i, q^j\rangle=0, \tag{1.22}$$

showing that the vectors q^i and q^j are orthogonal.

From (1.19) it is deduced that eigenvectors are determined up to a scalar multiple. In geometric language, only the orientations but not the lengths of the vectors are determined. It follows that the orthogonality (1.22), can be transformed into orthonormality by normalizing the eigenvectors. Thus, dividing the vectors q^i by $\langle q^i,\ q^i\rangle^{\frac{1}{2}}$, one has

$$u^i=\frac{q^i}{\langle q^i,\ q^i\rangle^{\frac{1}{2}}}$$

and hence

$$\langle u^i,\ u^j\rangle=\delta_{ij}. \tag{1.23}$$

Reduction to diagonal form

A real symmetric matrix A with distinct eigenvalues can be reduced to a diagonal form by means of an orthogonal transformation.

Consider the orthogonal matrix P having the normalized eigenvectors of A as its columns,

$$P=[u^1,\ u^2,\ldots,\ u^n]. \tag{1.24}$$

This matrix is, indeed, orthogonal since

$$P'P=I \qquad (\text{or } P'=P^{-1})$$

in view of (1.23). The matrix A is then orthogonally congruent (and similar) to a matrix Λ described by

$$\Lambda=P'AP. \tag{1.25}$$

The matrix Λ can be written as

$$\Lambda=\begin{bmatrix}\langle u^1,\ Au^1\rangle & \langle u^1, Au^2\rangle & \cdots\\ \langle u^2,\ Au^1\rangle & \langle u^2, Au^2\rangle & \cdots\\ & \cdots\cdots\cdots & \\ & \cdots\cdots\cdots & \end{bmatrix}$$

But the elements of Λ can be expressed as

$$\langle u^i,\ Au^j\rangle=\langle u^i,\ \lambda_j u^j\rangle=\lambda_j\langle u^i,\ u^j\rangle=\lambda_j\delta_{ij} \tag{1.26}$$

which shows that the matrix Λ is diagonal with the eigenvalues of A as its elements. It will be shown later that the restriction on the eigenvalues (to be distinct) can be removed and that *any* real symmetric matrix is orthogonally similar (congruent) to the matrix of its eigenvalues (see G. square matrices, *b*).

A real quadratic form can similarly be diagonalized by means of an orthogonal transformation. Consider the quadratic form

$$f=\langle x, Ax\rangle; \tag{1.27}$$

the orthogonal transformation $x=Py$, where P is defined by (1.24), transforms the quadratic form (1.27) into

$$f=\langle Py, APy\rangle=\langle y, P'APy\rangle=\langle y, \Lambda y\rangle \tag{1.28}$$

which can also be expressed as

$$f=\lambda_1 y_1^2+\lambda_2 y_2^2+\cdots+\lambda_r y_r^2 \tag{1.29}$$

where r is the number of nonzero eigenvalues.

In applications it may be required that two quadratic forms are diagonalized simultaneously. This is possible under two circumstances:

1. *The matrices A and B of the given quadratic forms commute*, i.e., AB = BA. An orthogonal matrix can then (and only then) be found such that both A and B are orthogonally similar (congruent) to diagonal matrices.

To show this, assume that A has distinct eigenvalues and that λ is one of them. From

$$Aq=\lambda q$$

one has

$$BAq=\lambda Bq,$$

or

$$ABq=\lambda Bq,$$

indicating that Bq is also an eigenvector. Since λ was assumed to be distinct, q and Bq must be a scalar multiple of each other. Hence

$$Bq=\mu q$$

where the scalar μ must then be an eigenvalue of B, and q is an eigenvector of B as well as A. Since this result holds for all eigenvalues, if A is diagonalized by

$$P'AP=\Lambda$$

as in (1.25) B will also be diagonalized simultaneously, the eigenvalues μ_i appearing as its diagonal elements.

It can be shown (see Ref [2]) that this result is valid even if the eigenvalues of A are not distinct.

2. *One of the two matrices is positive definite.* In this case the simultaneous reduction of the matrices (and hence the associated quadratic forms) can be achieved through a congruence transformation (not orthogonal).

The proof will this time be carried out directly on the quadratic forms. Consider the quadratic forms

$$f_1 = \langle x, Ax \rangle \quad \text{and} \quad f_2 = \langle x, Bx \rangle$$

and suppose f_1 is positive definite. Introduce the linear orthogonal transformation $x = Py$ to obtain

$$f_1 = \langle Py, APy \rangle = \langle y, P'APy \rangle = \langle y, \Lambda y \rangle$$

as is (1.28). Here all the diagonal elements λ_i are positive since A is positive definite, and

$$f_2 = \langle y, P'BPy \rangle.$$

Next, introduce

$$y = Qz$$

where

$$Q = \text{diag}[\lambda_1^{-\frac{1}{2}}, \lambda_2^{-\frac{1}{2}}, \ldots, \lambda_n^{-\frac{1}{2}}]$$

to obtain

$$f_1 = \langle z, Q'\Lambda Qz \rangle = \langle z, z \rangle$$

and

$$f_2 = \langle z, Cz \rangle \quad \text{where} \quad C = Q'P'BPQ.$$

Now introducing

$$z = Tv \quad \text{where} \quad T' = T^{-1}$$

such that C is diagonalized, one has

$$f_2 = \langle v, T'CTv \rangle = \mu_1 v_1^2 + \mu_2 v_2^2 + \cdots \tag{1.30}$$

and

$$f_1 = \langle v, v \rangle \tag{1.31}$$

Thus, the linear nonsingular transformation

$$x = PQTv$$

reduces the quadratic forms f_1 and f_2 to sums of squares simultaneously. Since

$$|T'Q'P'(B-\lambda A)PQT|=0$$

leads to

$$|\mu_i\delta_{ij}-\lambda I|=0,$$

the μ_i are the roots of $|B-\lambda A|=0$.

Before proceeding to discuss the canonical forms of general asymmetric matrices, two important properties of similarity transformations (which may have to be used instead of orthogonal transformations) must be noted:

1. Under similarity transformation the eigenvalues remain invariant. If the eigenvalues of the square matrix A are λ_i, the eigenvalues of

$$B=T^{-1}AT$$

are the same. Indeed,

$$|B-\lambda I|=|T^{-1}AT-\lambda I|=|T^{-1}(A-\lambda I)T|=|T^{-1}|\,|A-\lambda I|\,|T|=0$$

which yields

$$|A-\lambda I|=0 \quad \text{for} \quad |T|\neq 0.$$

The converse of this is not true; i.e., two matrices having the same eigenvalues may not be similar.

2. If $B=T^{-1}AT$ and p is the eigenvector of B corresponding to an eigenvalue λ, the eigenvector q of A corresponding to λ is $q=Tp$.

From

$$Bp=\lambda p$$

one has

$$T^{-1}ATp=\lambda p$$

or

$$ATp=\lambda Tp$$

which can be written as

$$Aq=\lambda q$$

giving

$$q=Tp. \tag{1.32}$$

General square matrices

Several important cases will be discussed:

(a) Every n-square matrix A is similar to a triangular matrix and the diagonal elements of the latter consist of the eigenvalues λ_i of A.

Suppose q^1 is an eigenvector corresponding to the eigenvalue λ_1. Form a nonsingular matrix T_l, the first column of which is q^1 while the remaining columns are arbitrary provided $|T_1| \neq 0$. Then, the first column of AT_1 is $Aq^1 = \lambda_1 q^1$, and the first column of $T_1^{-1}AT_1$ is $T_1^{-1}\lambda q^1$. This is, however, the first column of $T_1^{-1}\lambda_1 T_1$ which is $(\lambda_1, 0, \ldots, 0)'$. Hence,

$$T_1^{-1}AT_1 = \begin{bmatrix} \lambda_1 & B_1 \\ 0 & A_1 \end{bmatrix}$$

where A_1 is of order $n-1$. Since

$$|T_1^{-1}AT_1 - \lambda I| = (\lambda_1 - \lambda)\,|A_1 - \lambda I| = 0,$$

the eigenvalues of A_1 are $\lambda_2, \lambda_3, \ldots, \lambda_n$.

The procedure may now be repeated. Suppose q^2 is an eigenvector of A_1 corresponding to λ_2. Form a nonsingular matrix T_2 whose first column is q^2. Then,

$$T_2^{-2}A_1T_2 = \begin{bmatrix} \lambda_2 & B_2 \\ 0 & A_2 \end{bmatrix}$$

and so on. Finally, the matrix

$$T = T_1 \begin{bmatrix} I_1 & 0 \\ 0 & T_2 \end{bmatrix} \begin{bmatrix} I_2 & 0 \\ 0 & T_3 \end{bmatrix} \cdots \tag{1.33}$$

is obtained such that $T^{-1}AT$ is triangular with λ_i as its diagonal elements. Here I_k denotes a unit matrix of order k.

(b) If A is real with real eigenvalues, the above triangularization can be achieved through an orthogonal transformation.

After forming the nonsingular matrices T_1, T_2, etc., with q^1, q^2, etc. as their first columns, the Gram-Schmidt orthogonalization process can be used to transform these matrices into orthogonal matrices P_1, P_2, etc. whose first columns are proportional to q^1, q^2, etc., respectively. Since the product of orthogonal matrices is orthogonal, the resulting matrix T is orthogonal, and the real n-square matrix A with real eigenvalues is orthogonally similar to a triangular matrix, $P^{-1}AP$, whose diagonal elements are the eigenvalues of A.

Returning to the question which was left open earlier, it can now be shown readily that if A is real symmetric the triangular matrix $P^{-1}AP$ is, in fact, diagonal and this result is independent of coincidence of eigenvalues. Thus, note that if A is symmetric, so are $P_1'AP_1$, $P_2'AP_2$, etc. It follows that B_1, B_2, etc. are zero and the resulting matrix is therefore diagonal whose elements are the eigenvalues of A. This result was proved earlier under the restriction that the eigenvalues of A were distinct.

If the real n-square matrix A has complex eigenvalues, then the triangularization can be achieved by a unitary transformation. That is, there exists a unitary matrix U such that $U^{-1}AU = \bar{U}'AU$ is a triangular matrix whose diagonal elements are the eigenvalues of A. This result also holds for complex n-square matrices.

(c) If an n-square matrix A has n linearly independent eigenvectors, then it can be diagonalized by means of a similarity transformation.

Consider the nonsingular matrix

$$T = [q^1, q^2, \ldots, q^n]$$

having the linearly independent eigenvectors of A as its columns. Then,

$$AT = [Aq^1, Aq^2, \ldots, Aq^n]$$

and using $Aq^i = \lambda_i q^i$,

$$AT = [\lambda_1 q^1, \lambda_2 q^2, \ldots, \lambda_n q^n]$$

$$= [q^1, q^2, \ldots, q^n]\begin{bmatrix} \lambda_1 & 0 & \cdots & \\ 0 & \lambda_2 & \cdots & \\ \cdot & \cdot & \ddots & \\ & & & \lambda_n \end{bmatrix}$$

$$= T\Lambda$$

which yields

$$T^{-1}AT = \Lambda. \tag{1.34}$$

Thus, if the square matrix A has linearly independent eigenvectors, than it is similar to the diagonal matrix of its eigenvalues. *Conversely, if A is similar to a diagonal matrix, then it has a full complement of linearly independent eigenvectors.* To prove this, suppose A is similar to a diagonal matrix D,

$$T^{-1}AT = D.$$

Since the eigenvectors of a diagonal matrix are always linearly independent and the set of unit vectors e^i, satisfying $\langle e^i, e^j\rangle = \delta_{ij}$, can be taken as the eigenvectors of D, the eigenvectors q^i of A are given (1.32) by

$$q^i = Te^i.$$

But T is a nonsingular matrix and the q^i are therefore linearly independent.

(d) If the eigenvalues of an n-square matrix A are distinct, then A has n linearly independent eigenvectors.

Suppose the eigenvectors q^i of A are not linearly independent. Then, for some scalars α_i,

$$\alpha_1 q^1 \alpha_2 q^2 + \cdots + \alpha_n q^n = 0.$$

Pre-multiplying this equation by A and using $Aq^i = \lambda_i q^i$ results in

$$\alpha_1 \lambda_1 q^1 + \alpha_2 \lambda_2 q^2 + \cdots + \alpha_n \lambda_n q^n = 0.$$

Repeating the above process finally yields

$$\begin{bmatrix} 1 & 1 & \cdots & 1 \\ \lambda_1 & \lambda_2 & \cdots & \lambda_n \\ \lambda_1^2 & \lambda_2^2 & \cdots & \lambda_n^2 \\ \cdot & \cdot & & \cdot \\ \cdot & \cdot & & \cdot \\ \cdot & \cdot & & \cdot \\ \lambda_1^n & \lambda_2^n & \cdots & \lambda_n^n \end{bmatrix} \begin{bmatrix} \alpha_1 q^q \\ \alpha_2 q^2 \\ \cdot \\ \cdot \\ \cdot \\ \cdot \\ \alpha_n q^n \end{bmatrix} = 0$$

However, the determinant of the matrix on the left-hand side of the above equation is known as the Vandermondian which cannot be zero, indicating that all $\alpha_i = 0$. Hence, the eigenvectors q^i are linearly independent.

(e) A real $n \times n$ skew-symmetric matrix A can be reduced to the canonical form

$$T^{-1}AT = \text{diag}\left[\begin{pmatrix} 0 & \mu_1 \\ -\mu_1 & 0 \end{pmatrix}, \ldots, \begin{pmatrix} 0 & \mu_r \\ -\mu_r & 0 \end{pmatrix}, 0, \ldots, 0\right] \tag{1.35}$$

by means of an orthogonal transformation ($T^{-1} = T'$), where r is the number of non-zero eigenvalues.

It is first noted that the eigenvalues of a real skew-symmetric matrix A are either pure imaginary or zero, since in

$$\langle \bar{q}^i, Aq^i \rangle = \lambda_i \langle \bar{q}^i, q^i \rangle$$

the left-hand side was shown to be imaginary. Thus, assuming that $\lambda = i\mu$

is an eigenvalue, where μ is real, it can next be shown that the corresponding eigenvector $q = x + iy$, where x and y are real, is such that $\langle x, y\rangle = 0$. To this end, pre-multiply $Aq = i\mu q$ by q' and use $\langle q, Aq\rangle = 0$ to obtain

$$\langle q, Aq\rangle = i\mu\langle q, q\rangle = i\mu\langle x + iy, x + iy\rangle = 0$$

which yields

$$\langle x, y\rangle = 0.$$

Now using x and y as the first and second columns to form an orthogonal matrix $T_1 = [x, y, \ldots]$ whose remaining columns can be chosen arbitrarily provided $T_1'T_1 = I$, one has

$$T_1'AT_1 = \begin{bmatrix} \langle x, Ax\rangle & \langle x, Ay\rangle & \\ & & 0 \\ \langle y, Ax\rangle & \langle y, Ay\rangle & \\ & & \\ & & A_2 \end{bmatrix} = \begin{bmatrix} 0 & \mu & \\ & & 0 \\ -\mu & 0 & \\ & & \\ & & A_2 \end{bmatrix}$$

where A_2 is of order $n-2$ and is still skew-symmetric, the latter property being invariant under the congruence transformation. The last equality follows from

$$Aq = i\mu q$$

which upon pre-multiplying by x yields

$$\langle x, Ax\rangle + i\langle x, Ay\rangle = i\mu\langle x, x\rangle$$

resulting in

$$\langle x, Ay\rangle = \mu$$

and hence

$$\langle y, Ax\rangle = -\mu, \quad \text{since} \quad \langle x, x\rangle = 1.$$

As in the triangularization, this procedure can now be repeated for A_2 and then for the resulting A_3, etc., completing the canonical form (1.35). Note that if the order of the matrix is odd, one of the diagonal elements is zero in addition to blocks corresponding to zero eigenvalues.

(f) *Jordan canonical form:*
For an arbitrary n-square A there exists a nonsingular matrix T such

that one has the Jordan canonical form

$$T^{-1}AT=\begin{bmatrix} J_{n_1}(\lambda_1) & 0 & \cdots & 0 \\ 0 & J_{n_2}(\lambda_2) & \cdots & 0 \\ \vdots & & \ddots & \vdots \\ 0 & 0 & \cdots & J_{n_r}(\lambda_r) \end{bmatrix} \tag{1.36}$$

where the λ_i are the eigenvalues of A (distinct or coincident) and

$$n_1+n_2+\cdots+n_r=n$$

The J_{n_i} are called *Jordan blocks* whose main diagonals consist of the eigenvalues of A and the diagonal above each main diagonal consists of ones; e.g. if $n_2=3$,

$$J_{n_2}=\begin{bmatrix} \lambda_2 & 1 & 0 \\ 0 & \lambda_2 & 1 \\ 0 & 0 & \lambda_2 \end{bmatrix},$$

If $n_1=n_2=\cdots=n_r=1$, the Jordan canonical form becomes a diagonal matrix. The maximum number of eigenvectors corresponding to an eigenvalue is the number of Jordan blocks in which the latter appears. For example, if the matrix A is of order 3 and has the Jordan form

$$T^{-1}AT=\begin{bmatrix} \lambda_1 & 0 & 0 \\ 0 & \lambda_1 & 1 \\ 0 & 0 & \lambda_1 \end{bmatrix},$$

there exist 2 independent eigenvectors corresponding to λ_1. The number of Jordan blocks in which an eigenvalue appears is called the *index of the eigenvalue* and it is clear that the *multiplicity* of an eigenvalue is not less than its *index*. When the index is equal to multiplicity for all eigenvalues, then the eigenvectors of A are all linearly independent and the matrix is diagonalizable (or *simple* [3]).

The proof of (1.36) requires the introduction of certain additional concepts which are not necessary for the purpose of this chapter and it will not, therefore, be given (see Ref [1]).

(g) An n-square matrix A is called *normal* if the relation

$$A\bar{A}'=\bar{A}'A$$

holds. It can easily be seen that real symmetric, real skew-symmetric,

Hermitian, skew-Hermitian, diagonal, orthogonal and unitary matrices are normal. Also, if A is normal and U is unitary, the matrix $B=\bar{U}'AU$ is normal.

It can be shown that an n-square matrix A is unitarily similar to a diagonal matrix if and only if A is normal. Thus, note that under (b) of this section it was emphasized that any square matrix can be reduced to a triangular form by means of a unitary transformation; i.e., there exists a matrix $U^{-1}=\bar{U}'$ such that

$$B=\bar{U}'AU$$

is triangular. Since B is also normal ($B\bar{B}'=\bar{B}'B$) and the product of a scalar with its complex conjugate is always positive, all the off-diagonal elements of B are zero. On the other hand, if A is diagonal it is normal, showing that

$$B=\operatorname{diag}[\lambda_1, \lambda_2, \ldots, \lambda_n] \tag{1.37}$$

if and only if A is normal.

Exercises

1. Show that the eigenvalues of an Hermitian matrix are real and the eigenvectors associated with distinct eigenvalues are orthogonal in the sense that $\langle \bar{q}^i, q^j\rangle = 0$ for $i \neq j$.

2. Show that an Hermitian matrix can be reduced to a diagonal form.

3. If $H=A+iB$ is a positive definite Hermitian matrix, where A and B are real, then

$$|A| \geq |H|$$

with equality holding if and only if $B=0$.

4. Show that a necessary and sufficient set of conditions that a real symmetric matrix A is positive definite is that all the leading principal minors are positive:

$$a_{11}>0, \qquad \begin{vmatrix} a_{11} & a_{12} \\ a_{21} & a_{22} \end{vmatrix} > 0, \text{ etc.}$$

5. What is meant by the square root $A^{\frac{1}{2}}$ of a positive definite matrix A? Is it also positive definite? Why?

6. If A is real symmetric, then $I+\varepsilon A$ is positive definite for sufficiently small ε.

7. Show that

$$|A \pm B| > 0$$

where $A = A' > 0$ (p.d.) and $B = -B'$ (A and B are real).

1.4 Symmetrizable matrices

It has been shown by Taussky and Zassenhaus [4] that for every n-square matrix A there is a nonsingular symmetric matrix S transforming A into its transpose through

$$A' = S^{-1}AS \tag{1.38}$$

In this section, however, attention is focussed on real matrices, and writing (1.38) in the form

$$A = SA'S^{-1},$$

one observes that a real asymmetric matrix can be expressed as the product of two symmetric matrices [5] since both SA' and S^{-1} are symmetric. The symmetry of SA' follows from

$$SA' = AS.$$

The real matrix A will be called *symmetrizable* if the transformation matrix S can be chosen as a positive definite matrix; i.e. a real asymmetric matrix A is *symmetrizable* if it can be expressed as the product of two symmetric matrices one of which is positive definite.

It can be shown that a symmetrizable matrix is similar to a symmetric matrix via a positive definite transformation matrix.
Suppose

$$A = S_1 S_2 \tag{1.39}$$

where

$$S_1 = S_1', \qquad S_2 = S_2', \quad \text{and} \quad S_1 > 0 \text{ (p.d.)},$$

then

$$A = S_1^{\frac{1}{2}} S_1^{\frac{1}{2}} S_2$$

and

$$S_1^{-\frac{1}{2}} A S_1^{\frac{1}{2}} = S_1^{\frac{1}{2}} S_2 S_1^{\frac{1}{2}}$$

or

$$TAT^{-1} = K \tag{1.40}$$

where evidently $K = S_1^{\frac{1}{2}} S_2 S_1^{\frac{1}{2}}$ is symmetric and $T = S_1^{-\frac{1}{2}} > 0$.

Another interesting way of characterizing a symmetrizable matrix is based on

$$S_1^{-1}A = S_2 \tag{1.41}$$

which indicates that a symmetrizable matrix A becomes symmetric when multiplied by an appropriate p.d. matrix. Furthermore, since a symmetrizable matrix is similar to a symmetric matrix, its eigenvalues are necessarily real.

The symmetric matrix K in (1.40) can be diagonalized by means of an orthogonal matrix P, and it follows from (1.40) that *a symmetrizable matrix is similar to the diagonal matrix of its eigenvalues. Then, according to Section* 1.3(*c*), *eigenvectors of a symmetrizable matrix are linearly independent.* It can readily be demonstrated that the eigenvectors of a symmetrizable matrix have further interesting features.

In fact, denoting right and left eigenvectors of a symmetrizable n-square matrix A by q^i and r^j respectively, one has

$$\left.\begin{aligned} &(A-\lambda_i I)q^i = 0, \qquad i=1,2,\ldots,n \\ &\text{and} \\ &r^{i\prime}(A-\lambda_i I) = 0, \qquad i=1,2,\ldots,n. \end{aligned}\right\} \tag{1.42}$$

or using (1.40),

$$(T^{-1}KT-\lambda_i I)\, q^i = 0 \tag{1.43}$$

and

$$r^{i\prime}(T^{-1}KT-\lambda_i I) = 0. \tag{1.44}$$

Since K is symmetric, (1.43) and (1.44) lead to the relation

$$Tq^i = T^{-1}r^i$$

and hence

$$r^i = T^2q^i, \qquad i=1,2,\ldots,n. \tag{1.45}$$

It is further observed that, due to the symmetry of K, the inner product of Tq^i and Tq^j corresponding to distinct eigenvalues satisfies

$$\langle Tq^i, Tq^j\rangle = 0 \quad \text{for} \quad i \neq j$$

which leads to

$$\langle q^i, T^2q^j\rangle = 0 \quad \text{for} \quad i \neq j, \tag{1.46}$$

indicating orthogonality of the eigenvectors with respect to the positive definite matrix T^2. Similarly,

$$\langle r^i, (T^{-1})^2r^j\rangle = 0 \quad \text{for} \quad i \neq j. \tag{1.47}$$

Using (1.45), (1.47) yields the bi-orthogonality relation

$$\langle r^i, q^j \rangle = 0 \quad \text{for} \quad i \neq j \tag{1.48}$$

between the right and left eigenvectors corresponding to distinct eigenvalues.

If in (1.39) both S_1 and S_2 are positive definite, then the matrix K in (1.40) is also positive definite according to Sylvester's Law of inertia (which will be discussed in Section 1.6) and A has positive eigenvalues. This can be shown independently by observing that S_2 can be expressed as $S_2 = BB'$, and hence $K = S_1^{\frac{1}{2}} S_2 S_1^{\frac{1}{2}} = S_1^{\frac{1}{2}} BB' S_1^{\frac{1}{2}} = CC'$.

Exercises

1. Extend the concept of symmetrizability and its implications to complex matrices by using Hermitian matrices instead of real symmetric matrices.

2. Show that the eigenvalues of the symmetrizable matrix $A = S_1 S_2$ where $S_1 = S_1'$, $S_2 = S_2'$ and $S_1 > 0$ are given by $d_k = \theta_k \lambda_k$ in which λ_k are the eigenvalues of S_2 and $\theta_k > 0$ lies between the smallest and largest eigenvalues of S_1 (see Section 1.6).

1.5 Extremum properties of eigenvalues

Consider the matrices A and B of the same dimension and with elements from a field F (real or complex), and construct $A\lambda + B$ where λ is a scalar. The matrices obtained this way form a *pencil* of matrices [3]. A pencil of matrices A and B is *regular* if both A and B are square and A is nonsingular. The determinant

$$|A\lambda + B| = 0 \tag{1.49}$$

yields n roots for a regular pencil since A is nonsingular.

It is evident that (1.49) is related to eigenvalue problems, and with $-\lambda$, a regular matrix pencil may be associated with the eigenvalue problem

$$(A^{-1}B - \lambda I)q = 0.$$

In other words, the eigenvalues of the matrix $A^{-1}B$ are the roots associated with the regular matrix pencil $-A\lambda + B$, and such roots are often referred to as *eigenvalues* of the "generalized eigenvalue problem"

$$Bq = \lambda Aq. \tag{1.50}$$

A regular matrix pencil, $-A\lambda + B$, of order n is called *simple* [3] if it has n linearly independent right eigenvectors. It will be seen in the sequel that the latter condition actually implies the linear independence of left eigenvectors as well. It can further be shown that the right and left eigenvectors of a simple pencil corresponding to distinct eigenvalues are orthogonal with respect to A.
From

$$(-A\lambda_i + B)q^i = 0, \tag{1.51a}$$

$$r^{i\prime}(-A\lambda_i + B) = 0, \tag{1.51b}$$

and

$$(-A\lambda_j + B)q^j = 0, \tag{1.52a}$$

$$r^{j\prime}(-A\lambda_j + B) = 0, \tag{1.52b}$$

where $\lambda_i \neq \lambda_j$, one obtains

$$-\langle r^i, Aq^j\rangle\lambda_i + \langle r^i, Bq^j\rangle = 0 \tag{1.53}$$

and

$$-\langle r^i, Aq^j\rangle\lambda_j + \langle r^i, Bq^j\rangle = 0. \tag{1.54}$$

Substracting (1.54) from (1.53) results in

$$\langle r^i, Aq^j\rangle(\lambda_j - \lambda_i) = 0$$

or

$$\langle r^i, Aq^j\rangle = 0 \tag{1.55}$$

due to $\lambda_i \neq \lambda_j$. Since right (and left) eigenvectors are defined within an arbitrary constant multiplier, they can be normalized with respect to A, yielding the *biorthogonality* condition

$$\langle r^i, Aq^j\rangle = \delta_{ij} \tag{1.56}$$

which, by virtue of (1.53), gives

$$\langle r^i, Bq^j\rangle = \delta_{ij}\lambda_i. \tag{1.57}$$

The biorthogonality conditions (1.56) and (1.57) actually hold for all simple matrix pencils. In other words, a given simple matrix pencil can always be represented in the form [3]

$$R'AQ = I \quad \text{and} \quad R'BQ = \Lambda \tag{1.58}$$

where

$$Q=[q^1, q^2, \ldots, q^n],$$
$$R=[r^1, r^2, \ldots, r^n],$$
$$\Lambda=\operatorname{diag}[\lambda_1, \lambda_2, \ldots, \lambda_n],$$

and the matrix forms in (1.58) are equivalent to (1.56) and (1.57). To show this, write the n equations (1.51a) as

$$-AQ\Lambda+BQ=0 \tag{1.59}$$

and (1.51b) as

$$-\Lambda R'A+R'B=0 \tag{1.60}$$

From (1.59),

$$-Q\Lambda+A^{-1}BQ=0,$$

or

$$-\Lambda Q^{-1}+Q^{-1}A^{-1}B=0,$$

or

$$-\Lambda(AQ)^{-1}A+(AQ)^{-1}B=0,$$

which takes the form of (1.60) upon defining

$$R'=(AQ)^{-1}. \tag{1.61}$$

Hence R is the matrix of the left eigenvectors, and (1.58) follows from (1.61) and (1.59). It is also noted from (1.61) that the left eigenvectors of a simple matrix pencil are linearly independent as remarked earlier.

A variational description of eigenvalues can now be given on the basis of right and left eigenvectors. Thus, pre-multiplying (1.51a) by r^i the eigenvalues can be expressed as

$$\lambda_i=\frac{\langle r^i, Bq^i\rangle}{\langle r^i, Aq^i\rangle}, \qquad i=1,2,\ldots,n. \tag{1.62}$$

Replacing q^i and r^i in (1.62) by arbitrary vectors x and y, one obtains the *generalized Rayleigh quotient* [3, 6]

$$R(x, y)=\frac{\langle y, Bx\rangle}{\langle y, Ax\rangle}, \tag{1.63}$$

assuming the denominator is nonzero.

It can be shown that, for a simple matrix pencil whose right and left eigenvectors satisfy the relations (1.58), the quotient $R(x, y)$ has a *stationary* value at each pair of right and left eigenvectors, q^i and r^i, corresponding to an eigenvalue λ_i, and $R(q^i, r^i) = \lambda_i$. Indeed, considering the first variation of $R(x, y)$

$$\delta R(x, y) = \frac{(\langle \delta y, Bx\rangle + \langle y, B\delta x\rangle)\langle y, Ax\rangle - (\langle \delta y, Ax\rangle + \langle y, A\delta x\rangle)\langle y, Bx\rangle}{\langle y, Ax\rangle^2},$$

and evaluating at $x = q^i$, $y = r^i$ where

$$\langle \delta y, Bx\rangle = \lambda_i\langle \delta r^i, Aq^i\rangle,$$

$$\langle y, B\delta x\rangle = \langle \delta x, B'y\rangle = \lambda_i\langle \delta q^i, A'r^i\rangle,$$

$$\langle y, A\delta x\rangle = \langle \delta x, A'y\rangle = \langle \delta q^i, A'r^i\rangle$$

and

$$\langle y, Bx\rangle = \lambda_i\langle r^i, Aq^i\rangle$$

yield

$$\delta R(x, y) = 0 \quad \text{at} \quad x = q^i,\ y = r^i \tag{1.64}$$

since it can be assumed that $\langle r^i, Aq^i\rangle = 1$ for a given pair of q^i and r^i.

Also,

$$R(q^i, r^i) = \lambda_i, \qquad i = 1, 2, \ldots, n. \tag{1.65}$$

The *classical Rayleigh quotient* is obtained from (1.63) as a special case in which both A and B are real symmetric and A is positive definite,

$$R(x) = \frac{\langle x, Bx\rangle}{\langle x, Ax\rangle}. \tag{1.66}$$

The latter condition is significant since any real square matrix can be expressed as a product of two symmetric matrices as remarked earlier. If, in addition to symmetry properties, one of the matrices (say A) is positive definite, then $A^{-1}B$ is symmetrizable and it has linearly independent eigenvectors as well as other associated properties. It follows, in particular, that the matrix pencil associated with the classical Rayleigh quotient is necessarily *simple*, and the stationary property of the G.R.Q. (generalized Rayleigh quotient) applies.

It can be shown that the C.R.Q. has powerful maximum-minimum properties in addition to being stationary with respect to x at $x = q^i$. Consider

$$(-A\lambda_i + B)q^i = 0 \tag{1.67}$$

where $A = A'$, $B = B'$ and $A > 0$. Evidently, left and right eigenvectors are now identical for a given eigenvalue λ_i which can be written as

$$\lambda_i = \frac{\langle q^i, Bq^i\rangle}{\langle q^i, Aq^i\rangle}. \tag{1.68}$$

Suppose the eigenvalues (which are necessarily real) are in the ascending order

$$\lambda_1 \leq \lambda_2 \leq \cdots \leq \lambda_n. \tag{1.69}$$

An arbitrary real vector x in the space of eigenvectors can be expressed as

$$x = \alpha_i q^i \tag{1.70}$$

where the α_i are certain constants. Using

$$\langle q^i, Aq^j\rangle = \delta_{ij} \quad \text{and} \quad \langle q^i, Bq^j\rangle = \lambda_i\delta_{ij}, \tag{1.71}$$

the quotient (1.66) takes the form

$$\begin{aligned} R(x) &= \frac{\langle \alpha_i q^i, B\alpha_i q^i\rangle}{\langle \alpha_i q^i, A\alpha_i q^i\rangle} \\ &= \frac{\lambda_1\alpha_1^2 + \lambda_2\alpha_2^2 + \cdots + \lambda_n\alpha_n^2}{\alpha_1^2 + \alpha_2^2 + \cdots + \alpha_n^2}. \end{aligned} \tag{1.72}$$

Hence,

$$R(x) - \lambda_1 = \frac{(\lambda_2 - \lambda_1)\alpha_2^2 + (\lambda_3 - \lambda_1)\alpha_3^2 + \cdots + (\lambda_n - \lambda_1)\alpha_n^2}{\sum_i \alpha_i^2} \geq 0$$

and

$$R(x) - \lambda_n = \frac{(\lambda_1 - \lambda_n)\alpha_1^2 + (\lambda_2 - \lambda_n)\alpha_2^2 + \cdots + (\lambda_{n-1} - \lambda_n)\alpha_{n-1}^2}{\sum_i \alpha_i^2} \leq 0$$

which can be expressed as

$$\lambda_1 \leq \frac{\langle x, Bx\rangle}{\langle x, Ax\rangle} \leq \lambda_n.. \tag{1.73}$$

If x is also normalized with respect to A such that $\langle x, Ax\rangle = 1$, then the extremum property (1.73) takes the form

$$\lambda_1 \leq \langle x, Bx\rangle \leq \lambda_n. \tag{1.74}$$

This result shows that the minimum (maximum) value of the

Rayleigh quotient is the smallest (greatest) eigenvalue; i.e.,

$$\min_x R(x) = \lambda_1 \quad \text{and} \quad \max_x R(x) = \lambda_n. \tag{1.75}$$

Similarly, λ_2, λ_3, etc. can be given a variational characterization by noting that if the variable vector x is orthogonal to Aq^1, for example, the minimum value of $R(x)$ will be λ_2 [consider (1.72)]. In general, the rth eigenvalue λ_r is the minimum value of the function $R(x)$ for all vectors orthogonal to the first $r-1$ eigenvectors $q^1, q^2, \ldots, q^{r-1}$. Thus, one has

$$\left.\begin{aligned} \lambda_2 &= \min_{\langle x, Aq^1\rangle=0} \frac{\langle x, Bx\rangle}{\langle x, Ax\rangle}, \\ \lambda_3 &= \min_{\substack{\langle x, Aq^1\rangle=0 \\ \langle x, Aq^2\rangle=0}} \frac{\langle x, Bx\rangle}{\langle x, Ax\rangle}, \end{aligned}\right\} \tag{1.76}$$

etc.

Alternatively, if the constraints on x are selected such that x is orthogonal to the last $n-r$ eigenvectors, then λ_r is the maximum value of the function $R(x)$.

The variational characterization of an eigenvalue λ_r in the form of (1.76), however, involves $r-1$ preceding vectors and cannot therefore be used unless these vectors are known.

Courant-Fischer's max-min theorem

An independent variational characterization of eigenvalues can be formulated by introducing general constraints. Consider, for example, a constraint-vector y; the vectors x can then be required to satisfy the constraint $\langle x, y\rangle = 0$. In the context of the preceding discussion, the constraint vector y will be assumed to be given as $y = Az$. Then, *the rth eigenvalue is the maximum value which can be given to the minimum of $R(x)$ under $r-1$ constraints.* This statement can be written as follows:

$$\left.\begin{aligned} \lambda_1 &= \min_{R_n} \frac{\langle x, Bx\rangle}{\langle x, Ax\rangle} \\ \lambda_2 &= \max_{R_{n-1}} \min_{\langle x, Az^1\rangle=0} \frac{\langle x, Bx\rangle}{\langle x, Ax\rangle} \\ &\cdots\cdots\cdots\cdots\cdots \\ \lambda_r &= \max_{R_{n-r+1}} \min_{\substack{\langle x, Az^i\rangle=0 \\ (i=1,2,\ldots r-1)}} \frac{\langle x, Bx\rangle}{\langle x, Ax\rangle} \\ &\cdots\cdots\cdots\cdots\cdots \end{aligned}\right\} \tag{1.77}$$

Here min is taken over all vectors x in the subspace R_{n-r+1} of dimension $n-r+1$, and max over all subspaces R_{n-r+1}.

The following self-evident principle will be useful in proving the above theorem: If certain constraints are added to a problem, the corresponding minimum cannot be lower than the minimum of the original problem since the set of functions considered in the constrained problem is, in fact, contained in the set of the original problem.

Consider now the $r-1$ independent constraints

$$\langle x, Az^i\rangle = 0 \qquad i=1,2,\ldots,r-1, \tag{1.78}$$

and introduce the special conditions

$$\langle x, Aq^j\rangle = 0 \qquad j=r+1,\ldots,n. \tag{1.79}$$

In the n dimensional space of eigenvectors there exists a vector $\hat{x}$ which satisfies these $n-1$ constraints. Since $\hat{x}$ is orthogonal to Aq^j $(j=r+1,\ldots,n)$ and by virtue of (1.76) and its alternate,

$$\frac{\langle \hat{x}, B\hat{x}\rangle}{\langle \hat{x}, A\hat{x}\rangle} \leq \lambda_r.$$

But the variable vector x has less constraints than $\hat{x}$ and one therefore has

$$\min_{\substack{\langle x, Az^i\rangle = 0\\ (i=1,2,\ldots,r-1)}} \frac{\langle x, Bx\rangle}{\langle x, Ax\rangle} \leq \frac{\langle \hat{x}, B\hat{x}\rangle}{\langle \hat{x}, A\hat{x}\rangle} \leq \lambda_r. \tag{1.80}$$

Since (1.80) holds for all variable constraints (1.78) it implies that

$$\lambda_r = \max_{R_{n-r+1}} \quad \min_{\substack{\langle x, Az^i\rangle = 0\\ (i=1,2,\ldots,r-1)}} \frac{\langle x, Bx\rangle}{\langle x, Ax\rangle}. \tag{1.81}$$

The max is attained when the constraints (1.78) are chosen as the special constraints $\langle x, Aq^i\rangle = 0$, $i=1,2,\ldots,r-1$.

Exercises

1. Show that the smallest eigenvalue λ_1 of $(A+B)$ is greater than or equal to λ_1 (of A) $+\lambda_1$ (of B) if A and B are real symmetric positive definite matrices. If B is positive semi-definite, what can be said about the eigenvalues?

2. Extend the max-min properties of eigenvalues associated with simple pencils to Hermitian matrix pencils.

3. Show that

$$\lambda_1 \leq \langle \bar{x}, Ax \rangle \leq \lambda_n$$

where A is a real symmetric matrix with the eigenvalues

$$\lambda_1 \leq \lambda_2 \leq \cdots \leq \lambda_n$$

and x is a complex vector satisfying $\langle \bar{x}, x \rangle = 1$.

1.6 Sylvester's law of inertia and its quantitative formulation

Consider a real symmetric matrix A of order n and rank r. Let p be the number of positive eigenvalues; the number p is the *index* of the matrix A and the associated quadratic forms. The difference between the number of positive and negative eigenvalues, $p-(r-p)$, is called *signature.*

Sylvester's law of inertia asserts that r and p are invariant under congruence transformations; i.e., if Q is a real nonsingular matrix of order n, then

$$Q'AQ = D \equiv \operatorname{diag}[d_1, d_2, \ldots, d_n] \tag{1.82}$$

has $n-r$ elements d_i equal to zero and p positive elements.

Evidently, the rank of D is the same as that of A since Q is nonsingular. To show the latter part of the inertia law, suppose the d_i are in the descending order

$$d_1 \geq d_2 \geq \cdots \geq d_n, \tag{1.83}$$

and assume that the first x of them are positive; i.e., $d_1, d_2, \ldots, d_x > 0$ and $d_{x+1} \leq 0$. It is to be shown that $x = p$. The eigenvalues λ_i of A are given by

$$T'AT = \Lambda \equiv \operatorname{diag}[\lambda_1, \lambda_2, \ldots, \lambda_n] \tag{1.84}$$

where T is orthogonal and λ_i are assumed to be in the descending order

$$\lambda_1 \geq \lambda_2 \geq \cdots \geq \lambda_n. \tag{1.85}$$

Since the index of A is p, one has

$$\lambda_1, \lambda_2, \cdots, \lambda_p > 0 \quad \text{and} \quad \lambda_{p+1} \leq 0.$$

Consider now the subspace R_1 defined by the first p columns t_i of T, and the subspace R_2 defined by the last $n-x$ columns q_j ($j = x+1, \ldots, n$) of Q. An arbitrary vector y in R_1 can be expressed as

$$y = \alpha_i t_i, \qquad i = 1, 2, \ldots, p,$$

and one has

$$\langle y, Ay\rangle = \lambda_i \alpha_i^2 > 0.$$

On the other hand, an arbitrary vector z in R_2 of the form

$$z = \beta_i q_i, \qquad i = x+1, \ldots, n.$$

results in

$$\langle z, Az\rangle = d_i \beta_i^2 \leq 0, \qquad i = x+1, \ldots, n.$$

The dimension of the space $R_1 + R_2$ is given by $p + n - x$, and if $p > x$, a vector h can be found which is in both R_1 and R_2. But, for the vector h, one then has $\langle h, Ah\rangle > 0$ and $\langle h, Ah\rangle \leq 0$ which is not possible. It follows that p cannot be larger than x. Considering the spaces spanned by the sets $t_{p+1}, \ldots, t_n$ and $q_1, \ldots, q_x$, it can similarly be demonstrated that x cannot be larger than p. It is then concluded that $x = p$.

Sylvester's law of inertia, as discussed above, provides a qualitative link between the eigenvalues (1.84) of a real symmetric matrix A and those of $Q'AQ$ given in (1.82) such that the number of vanishing, positive and negative elements are the same in both sets. An interesting quantitative sharpening of this law has been presented by Ostrowski [7] who showed that the two sets of eigenvalues given in (1.82) and (1.84) are, in fact, linked by the relation

$$d_k = \theta_k \lambda_k, \qquad k = 1, 2, \ldots, n, \tag{1.86}$$

where θ_k lie between the smallest eigenvalue μ_1 and the largest eigenvalue μ_n of the positive definite matrix $Q'Q$.

In the special case when A is taken as the unity matrix I, it is seen that $d_k = \theta_k$ as expected. Also, if the transformation matrix T is orthogonal such that $Q' = Q^{-1}$, $d_k = \lambda_k$ since $\theta_k = 1$.

To show the relation (1.86), note that if the kth eigenvalue of A is λ_k, the kth eigenvalue of $A - \lambda_k I$ is zero, and according to Sylvester's law of inertia the kth eigenvalue of

$$Q'(A - \lambda_k I)Q = Q'AQ - \lambda_k Q'Q \tag{1.87}$$

is also zero. Now applying Courant-Fischer's theorem to the kth eigenvalue of (1.87), and assuming that d_i and λ_i are in ascending order, one has

$$0 = \max_{R_{n-k+1}} \min_{\substack{\langle x, x\rangle = 1 \\ x \in R_{n-k+1}}} [\langle x, Q'AQx\rangle - \lambda_k \langle x, Q'Qx\rangle] \tag{1.88}$$

which yields to (1.86).

An interesting result concerning symmetrizable matrices follows immediately. Thus, if $A = S_1 S_2$ is a symmetrizable matrix where $S_1 = S_1'$,

$S_2 = S_2'$ and $S_1 > 0$, and if the eigenvalues of S_2, S_1 and A are denoted by λ_i, μ_i and d_i, respectively, then the relation (1.86) holds where θ_k lie between μ_1 and μ_n. This is true because S_1 can always be expressed as QQ', and

$$Q^{-1}AQ = Q^{-1}QQ'S_2Q = Q'S_2Q. \tag{1.89}$$

In other words, the eigenvalues of a symmetrizable matrix are obtained by multiplying the corresponding eigenvalues of the symmetric matrix S_2 by certain positive constants which lie between the smallest and largest eigenvalues of the positive definite matrix S_1.

Exercises

1. Extend the concepts of this section to Hermitian matrices.

2. If H_1 and H_2 are two Hermitian matrices of order n, if the eigenvalues of H_1 and $H_1 - H_2$ ordered increasingly are λ_k, d_k $(k = 1, 2, \ldots, n)$ respectively, and if H_2 is positive definite, then

$$d_k = \lambda_k - h_k$$

where h_k is positive and lies between the smallest and largest eigenvalues of H_2.

2

Fundamental concepts of stability

2.1 Introductory remarks

Kelvin remarked on the importance and elusiveness of a concept of stability. Indeed, there is no universal definition of this important concept and it may be required to adopt an appropriate definition according to the nature of a particular problem at hand. Nevertheless, all the definitions of stability are basically concerned with the response of a system to certain disturbances. Considering a given state of motion or equilibrium of a system, it is often essential to know how sensitive the state is to perturbations. If the deviation of the perturbed motion from the original state remains within acceptable limits, then the state may be considered stable. It is, however, at this point where the questions concerning the nature and magnitude of the deviations and disturbances with regard to acceptable stability limits come into the picture. A metric (or several metrics) may be prescribed to measure the deviations from a given state together with acceptable limits for stability. Similarly, a measure(s) for disturbances may be introduced, such that if the deviations from the state under consideration remain within the limits prescribed for the metric(s) when the disturbances do not exceed the established measure(s) the state is called stable. Otherwise the state is unstable. Naturally, as far as the metrics and prescribed limits are concerned, what may be considered appropriate for a problem or set of problems is not necessarily so for other problems. Thus, a variety of definitions is laid down in different contexts. It is not impossible, however, to unify the concepts and definitions of stability in certain instances. Consider, for example, Lyapunov's original definition for discrete mechanical systems which stipulates that a state is stable whenever in the motion following a sufficiently small initial disturbance (in terms of initial displacements and velocities), the displacements and velocities remain as small as desired for all positive time. A direct application of this definition to continuous elastic bodies might lead to discrepancies [8]. Koiter [9] observed, however, that it is not physically meaningful to require that the displacements and velocities following a sufficiently small disturbance remain as small as desired all over the body.

Instead, he introduced L_2-norms of the displacement and velocity fields as metrics with respect to which the stability of a continuous body is defined [9, 10]. Koiter's modified definition of stability reduces to Lyapunov's original definition in the case of discrete systems. Thus, one has a single definition of stability for both discrete and continuous elastic systems, which is most desirable.

Consider a system whose behaviour can be described by a finite set of variables q_i $(i = 1, 2, \ldots, n)$, or generalized coordinates, and suppose the system is under the influence of m independent parameters η_j $(j = 1, 2, \ldots, m)$. Interest then focusses on the behaviour of the system as the parameters are varied in a slowly evolving process. Starting from an initially stable state, for example, will the system lose its stability at one stage? When and how? What can be said about the stability boundary and post-critical behaviour? The answers to these and similar questions may be provided in different ways. First of all, if the system under consideration is not a *gradient system,* that is, if it cannot be described by an all-embracing *potential,* then, despite the effectively statical process in which the control parameters η_j are varied, the system may lose its stability dynamically, involving oscillations with increasing amplitude. Such systems require a dynamical formulation and analysis. On the other hand, if the system is a gradient system possessing a potential (e.g., the total potential energy of conservative mechanical systems), statical criteria such as the *energy criterion* of Mechanics may be adopted for the analysis. In fact, a general nonlinear theory concerning the stability of multiple-parameter gradient systems has been developed over the years and has recently been presented as a monograph [10]. This theory identifies the distinct forms of equilibrium surfaces in the $m + n$ dimensional configuration space associated with the instabilities of the system, and provides general methods of nonlinear stability analysis as well as exploring quantitative and qualitative response characteristics of the system, including the stability boundary, post-critical behaviour, etc. Interestingly, Catastrophe Theory [11] developed in the context of Topology parallels this approach and gives a full but *qualitative* description of the phenomena which may be expected under certain circumstances. A discussion concerning the analogy and distinction between the two theories is presented in Ref. [12], together with the comparative advantages and/or disadvantages of both theories.

The main goal of this book is to present a general linear theory of stability concerning finite dimensional (discrete), multiple-parameter autonomous systems, both gradient and nongradient systems being treated from a dynamics point of view. This will be done by further classifying the systems into certain well-defined groups, and in this classification mostly

Ziegler's approach is followed [13]. Although most of the examples are chosen from Mechanics, the theory presented is applicable to problems in other disciplines, providing the essential variables and parameters are interpreted appropriately.

2.2 Lyapunov's definition of stability

Consider an autonomous holonomic dynamical system described by the generalized coordinates q_i $(i = 1, 2, \ldots, n)$ and independent parameters η_j $(j = 1, 2, \ldots, m)$. In the Lagrangian formulation, the free motion of this system is governed by n ordinary differential equations of second order. Alternatively, in the state space, the motion can be described by a set of first order differential equations. For a given set of η_j, the latter equations can be expressed in the vector form

$$\dot{x} = f(x) \tag{2.1}$$

where x is the state vector of order $2n$, having the generalized coordinates q_i and their time derivatives $\dot{q}_i$ as its elements, and $\dot{x} \equiv \mathrm{d}x/\mathrm{d}t$. An Euclidean space of $2n$ dimensions spanned by the coordinates x_i can be introduced to represent the motion of the system geometrically. This space is known as the *phase space* or *state space*, and an equilibrium state x^e satisfies $f(x^e) = 0$ for all t.

Without loss of generality, suppose $x = 0$ is an equilibrium state and that it's stability is to be investigated. According to Lyapunov, the state $q_i = \dot{q}_i = 0$ (that is, $x = 0$) is stable if $\|q_i(t)\|$ and $\|\dot{q}_i(t)\|$ remain arbitrarily small for all $t > 0$ provided the initial disturbance is such that $\|q_{io}\|$ and $\|\dot{q}_{io}\|$ are sufficiently small, where $q_{io} = q_i(0)$ and $\dot{q}_{io} = \dot{q}_i(0)$. Here it has to be noted that although the motivation for the state equation (2.1) comes from the n-degree-of-freedom system; nevertheless, the discussion based on (2.1) applies equally to all systems which can be described by differential equations of the type (2.1). If the dimension $2n$ for x is essential it will be emphasized.

In the geometric terms of the state space the motion and configuration of the system are represented by a point of this space and the stability of the origin can be defined with respect to the distance of this variable point from the origin of the space. Then the origin $x = 0$ is *stable* if it is possible to find a positive $\delta(\epsilon)$ for any sufficiently small $\epsilon > 0$ such that, following the initial disturbance

$$\|x^\circ\| < \delta, \quad \text{where} \quad x^\circ = x(0), \tag{2.2}$$

the inequality

$$\|x(t)\| < \epsilon \tag{2.3}$$

is satisfied for all $t > 0$. Otherwise the state $x = 0$ is unstable.

The state $x = 0$ is said to be *asymptotically stable* if it is stable and, in addition, one has

$$\lim_{t \to \infty} \|x(t)\| = 0. \tag{2.4}$$

In other words, the equilibrium state represented by the origin of the state space is stable if the variable point $x(t)$, representing the motion of the system in the state space, remains within a hypersphere of arbitrarily small radius (ϵ) provided the initial disturbance is such that the point x° is within a hypersphere of sufficiently small radius (δ). A two-dimensional representation is depicted in Figure 2.1. Similarly, the origin is asymptotically stable if the point $x(t)$ tends to origin as time becomes infinity. Instability occurs if the trajectory of the point $x(t)$ crosses the hypersphere of radius ϵ.

It is inferred from (2.1) that the definitions of stability and asymptotic stability given above are not restricted to linear systems although they are related to local properties. There are other definitions concerning global properties of the system, nonautonomous systems, motion of free as well as forced systems, etc. For the purpose of this book, however, Lyapunov's original definition is adequate and will be implied throughout.

If the solutions of the vector differential equation (2.1) or second order equations arising in the Lagrangian formulation can be obtained in closed form, one can establish certain *stability criteria* appealing directly

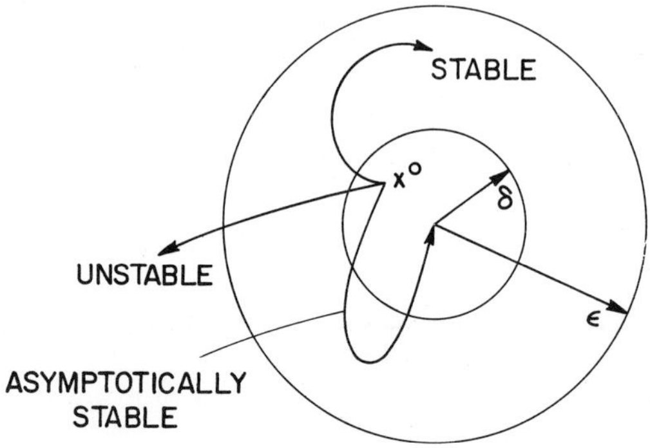

Figure 2.1

to the definitions, as will be discussed later. It is often desired, however, to obtain information concerning the stability properties of a system without actually solving the differential equations. This is particularly so when the solution of the differential equations poses serious difficulties or no closed form solution is obtainable. *Lyapunov's direct (second) method* is primarily aimed at circumventing these difficulties, and is developed on the model of Lagrange's stability theorem which employs a potential function to ascertain whether a state of equilibrium is stable or not. Lagrange's theorem is, of course, related to gradient systems only and asserts that *an equilibrium state is stable if the total potential energy at that point has a complete relative minimum* [10]. This idea has been generalized by Lyapunov and will now be described.

Let $V(x)$ be a real scalar function of the vector x, and assume that it is continuous with its first partial derivatives in a certain region R containing the origin. $V(x)$ is said to be a *Lyapunov function* if it is positive definite in R while its time derivative $\dot{V}(x)$ is negative semi-definite in R. Here the positive definiteness is defined as in Chapter 1, and it follows that the origin of the state space is an isolated minimum of $V(x)$. Also, note that

$$\dot{V} = \frac{\partial V}{\partial x_i} \dot{x}_i = \text{grad } V \cdot \dot{x} = \text{grad } V \cdot f(x). \tag{2.5}$$

The following two theorems have been formulated by Lyapunov:

a. *Stability Theorem.* The null solution of the system (2.1), i.e., the origin of the state space, is stable if in some region R about the origin there exists a Lyapunov function $V(x)$.
b. *Asymptotic Stability Theorem.* The origin is asymptotically stable if the derivative $\dot{V}(x)$ of the Lyapunov function is negative definite.

To prove these theorems, assume that the positive definite function $V = V(x)$ describes a cup-shaped surface in $V-x$ space (Fig. 2.2) such that its cross sections with $V=k$ planes form a set of curves enclosing each other in the state space. Referring now to the stability definition given earlier, one selects k such that the curve $V(x)=k$ is enclosed by the sphere of radius ϵ tangentially. It is then possible to choose δ such that the sphere of radius δ is just enclosed by $V(x)=k$, both touching each other tangentially. Since $V(x)<k$ for any initial point x° in the sphere of radius δ, any path with the initial state x° cannot reach $V(x)=k$ due to $\dot{V}\leq 0$. Hence, the path remains within the sphere of radius ϵ. If $\dot{V}<0$, $V(x)$ actually decreases along the path, but since it cannot be zero at any point other than the origin (V is

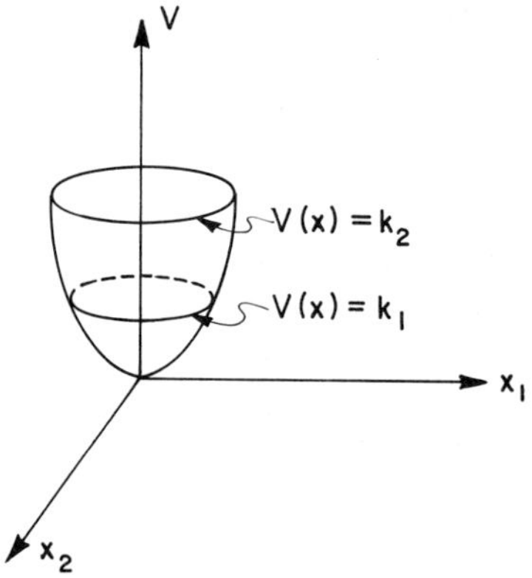

Figure 2.2

positive definite), the path has to tend to the origin as $V(x)$ tends to zero. At the origin $V(x)=0$.

These theorems provide sufficient conditions for stability. In other words, if one cannot find a Lyapunov function, it does not necessarily follow that the system is unstable. But there are no general systematic methods for finding Lyapunov functions either, and Lyapunov formulated a number of instability theorems which may rule out stability and enable one to avoid a waste of effort trying to construct Lyapunov functions. These theorems can be found in Refs. [14, 15, 16] and will not be given here. A thorough phase space analysis and a discussion on limit cycles as well as mathematical approximation methods can be found in Leipholz's book [15] which also contains illustrative examples.

2.3 Linear systems and stability criteria

Let the system described by (2.1) be a linear system; then (2.1) takes the form

$$\dot{x} = Ax \tag{2.6}$$

where A is a square matrix of order $2n$.

It can be shown that a necessary and sufficient condition for the null solution $x=0$ of (2.6) to be asymptotically stable is that all the eigenvalues of A have negative real parts. In other words, regardless of the value of $x(0)=x^{\circ}$, the solution of (2.6) approaches zero as t tends to infinity if and only if all the eigenvalues of A have negative real parts.

For convenience, a matrix A will be called *stable* (or *a* stability matrix) if all its eigenvalues have negative real parts. Thus, it must be shown that if A in (2.6) is a *stable matrix*, then the system (2.6) is asymptotically stable.

The solution of (2.6) can be obtained conveniently by making use of the canonical properties of general matrices discussed in Section 1.3. If, for example, A is similar to a diagonal matrix, then one can introduce the transformation

$$x = Ty, \qquad |T| \neq 0 \tag{2.7}$$

such that the resulting system

$$\dot{y} = T^{-1}ATy \tag{2.8}$$

is associated with the diagonal matrix

$$T^{-1}AT = \begin{bmatrix} \lambda_1 & 0 & & \\ 0 & \lambda_2 & & \\ & & \ddots & \\ & & & \lambda_{2n} \end{bmatrix} \tag{2.9}$$

where the λ_i are the eigenvalues of A. Note that, if the initial condition was given by $x(0) = x^\circ$, it is now in the form $y(0) = T^{-1}x^\circ \equiv y^\circ$, and the system (2.8) is in a canonical form which uncouples into $2n$ scalar equations

$$\dot{y}_i = \lambda_i y_i, \qquad y_i(0) = y_{io}, \tag{2.10}$$

yielding the well-known solutions

$$y_i = y_{io}e^{\lambda_i t}, \qquad i = 1, 2, \ldots, 2n. \tag{2.11}$$

Clearly, if $\text{Re}\,(\lambda_i) < 0$ for all roots, then

$$y_i \to 0 \quad \text{as} \quad t \to \infty.$$

In general, of course, A may not be diagonalizable; it was shown in Chapter 1, however, that every square matrix A can be transformed into a triangular form (Section 1.3(a) and (b)). Thus, if the transformation (2.7) is chosen appropriately, the matrix (2.9) takes a triangular form, and one has the scalar equations

$$\begin{aligned}
\dot{y}_1 &= b_{11}y_1 + b_{12}y_2 + b_{13}y_3 + \cdots + b_{1,2n}y_{2n} \\
\dot{y}_2 &= \qquad\quad\;\; b_{22}y_2 + b_{23}y_3 + \cdots + b_{2,2n}y_{2n} \\
&\cdot \quad \cdot \quad \cdot \quad \cdot \quad \cdot \quad \cdot \quad \cdot \quad \cdot \\
\dot{y}_{2n} &= \qquad\qquad\qquad\qquad\qquad\quad b_{2n,2n}y_{2n}
\end{aligned}$$

where b_{ij} are the elements of the triangular matrix $B = T^{-1}AT$ and the diagonal elements $b_{ii} = \lambda_i$ with $\mathrm{Re}(\lambda_i) < 0$ since A was assumed to be stable.

The last equation immediately yields the solution

$$y_{2n} = y_{2n,0}e^{\lambda_{2n}t}, \qquad y_{2n,0} \equiv y_{2n}(0) \tag{2.12}$$

which tends to zero as $t \to \infty$.

Proceeding successively, the next differential equation (from last) is

$$\dot{y}_{2n-1} = \lambda_{2n-1}y_{2n-1} + b_{2n-1,2n}y_{2n}$$

which is readily recognized as a linear first order equation. Using (2.12), the solution of this equation can be expressed as

$$y_{2n-1}e^{-\int\lambda_{2n-1}\mathrm{d}t} = \int b_{2n-1,2n}y_{2n,0}e^{(\lambda_n - \lambda_{2n-1})t}\,\mathrm{d}t + c.$$

If

$$\lambda_n \neq \lambda_{2n-1},$$

the solution takes the form

$$y_{2n-1} = ce^{\lambda_{2n-1}t} + \frac{b_{2n-1,2n}y_{2n,0}}{\lambda_{2n} - \lambda_{2n-1}}e^{\lambda_{2n}t} \tag{2.13}$$

where c is a constant. In the case $\lambda_{2n} = \lambda_{2n-1}$, one has

$$y_{2n-1} = y_{2n-1,0}e^{\lambda_{2n-1}t} + b_{2n-1,2n}y_{2n,0}te^{\lambda_{2n-1}t}. \tag{2.14}$$

Clearly, both (2.13) and (2.14) tend to zero as $t \to \infty$, since

$$\lim_{t\to\infty} te^{\lambda_{2n-1}t} = 0.$$

It is noted in passing that if $b_{2n-1,2n} = 0$, the solution (2.14) will not contain a term with t. Observing further that this condition also indicates linearly independent eigenvectors corresponding to $\lambda_{2n} = \lambda_{2n-1}$, one concludes that in the case of multiple roots, terms with t do not appear if the corresponding eigenvectors remain linearly independent.

The results for $y_{2n-2}, \ldots, y_1$ follow similarly.

Instead of the triangular form of A, one can use the Jordan canonical form (Section 1.3(f)) to prove the same result. Recalling the expansion

$$e^{At} = I + At + \cdots + \frac{A^n t^n}{n!} + \cdots \tag{2.15}$$

it is observed that

$$e^{At} = Te^{Jt}T^{-1} \tag{2.16}$$

where $J = T^{-1}AT$ is the Jordan canonical form of A. Also, on the basis of (2.15) one has

$$e^{Jt} = \text{diag}\,[e^{J_{n_1}t}, e^{J_{n_2}t}, \ldots, e^{J_{n_r}t}], \tag{2.17}$$

and

$$e^{J_{n_i}t} = \begin{bmatrix} 1 & t & t^2/2! & \cdots & t^{s-1}/(s-1)! \\ 0 & 1 & t & \cdots & t^{s-2}/(s-2)! \\ \cdot & \cdot & \cdot & \cdots & \cdots\cdots \\ 0 & 0 & 0 & \cdots & 1 \end{bmatrix} e^{\lambda_i t} \tag{2.18}$$

where J_{n_i} $(i = 1, 2, \ldots, r)$ are the Jordan blocks, and s is the order of J_{n_i}.

It can be verified that the solution of (2.6) may be expressed in the matrix form

$$x = e^{At}x^\circ, \qquad x^\circ = x(0), \tag{2.19}$$

and hence, by virtue of (2.16),

$$x = Te^{Jt}T^{-1}x^\circ. \tag{2.20}$$

From (2.17) and (2.18), it is deduced that the solution (2.20) tends to zero as $t \to \infty$ if A is a stable matrix, since

$$\lim_{t\to\infty} t^k e^{\lambda t} = 0$$

for any integer k. The converse of this follows easily.

It is noted that the solution can also be expressed in the transformed coordinates by introducing

$$x = Ty, \qquad |T| \neq 0$$

such that

$$\dot{y} = T^{-1}ATy = Jy, \qquad T^{-1}x^\circ = y^\circ,$$

where J is the Jordan canonical form. The solution is in the form

$$y = e^{Jt}y^\circ$$

where e^{Jt} is given by (2.17). Obviously, a reversed transformation leads to (2.20).

If at least one of the eigenvalues has positive real part, it is clear from (2.18) that the solution will be unbounded, and hence the system will be unstable.

If the matrix A has no eigenvalues with positive real parts but some eigenvalues have zero real parts, then asymptotic stability is ruled out but not stability. If, for example, the eigenvalues with zero real parts correspond to Jordan blocks of order one, it is easy to see that the solutions will be bounded. On the other hand, if the eigenvalues with zero real parts correspond to Jordan blocks of order larger than one, the solutions increase like powers of t, rendering the system unstable.

From Chapter 1 it is known that when the *index of all eigenvalues* is equal to their *multiplicity*, the eigenvectors of A are all linearly independent and the matrix A is of *simple* structure, the corresponding Jordan canonical form actually being diagonal. It is important, therefore, to bear in mind that it is not the multiplicity of eigenvalues alone that determines stability but rather the dimension of the eigenvector space. If all eigenvectors are linearly independent, then the multiplicity of eigenvalues becomes irrelevant as far as stability is concerned. On the other hand, a reduction in the dimension of the eigenvector space is always associated with a corresponding multiplicity of eigenvalues, resulting in a Jordan block of order larger than one.

It must be further noted that the above arguments apply to the linear system (2.6) when the latter is considered on its own. In practice, however, the linear system (2.6) is often a linearized model of a nonlinear system whose stability properties, in the case when at least one eigenvalue has zero real part while the remaining eigenvalues have negative real parts, may not coincide with those of the linearized system. It is, therefore, safer to call such cases *critical*, implying that higher order terms may be necessary to decide on stability. This point will be taken up again when the characteristic equation of second order differential equations is considered, and the meaning of 'criticality' in other contexts will be discussed.

The asymptotic behaviour of the solutions of (2.6) can be studied by making use of quadratic forms. This method was considered by Lyapunov.

Let A and $x(0)=x^\circ$ be real for the system

$$\dot{x}=Ax, \qquad x(0)=x^\circ,$$

and consider the quadratic form

$$V(x)=\langle x, Qx\rangle \tag{2.21}$$

where Q is a symmetric constant matrix. Then, one has

$$\begin{aligned}\dot{V}&=\langle \dot{x}, Qx\rangle+\langle x, Q\dot{x}\rangle\\ &=\langle Ax, Qx\rangle+\langle x, QAx\rangle\\ &=\langle x, (A'Q+QA)x\rangle\end{aligned}$$

where $A'Q+QA$ is necessarily symmetric.

If a positive definite matrix Q can be found such that

$$A'Q+QA=P \tag{2.22}$$

where P is an arbitrary negative definite matrix, the system is asymptotically stable and A is a stable matrix. It can be shown [2] that a unique positive definite matrix Q exists if and only if A is stable. If an indefinite or negative matrix Q is found, then the equilibrium is unstable. This is, however, a sufficient condition only. It is not difficult to show [17] that (2.22) is equivalent to

$$A'Q+QA=-I. \tag{2.23}$$

For, $P=-T'T$ by virtue of its definiteness, and one has

$$A'Q+QA=-T'T$$

or

$$(TAT^{-1})'\cdot(T')^{-1}QT^{-1}+(T')^{-1}QT^{-1}\cdot TAT^{-1}=-I$$

which can be written as

$$B'S+SB=-I \tag{2.24}$$

where $B=TAT^{-1}$ can only be stable if A is (and vice versa) and $S=(T')^{-1}QT^{-1}$ is positive definite if and only if Q is.

On the other hand, it is interesting to see that Lyapunov's equation (2.22) is actually equivalent to Stein's theorem [17] which asserts that if the eigenvalues of a matrix C lie inside the unit circle, then there exists a positive definite matrix Q such that

$$Q-C'QC \tag{2.25}$$

is positive definite. Here C is assumed to be real.

To show the equivalence, one uses the Cayley transformation which is an extension of the transformation $\beta=(1+\alpha)(1-\alpha)^{-1}$ carrying the complex number α with negative real part into a number β inside the unit circle. Thus, the stable matrix A can be transformed [17] into

$$C=(I+A)(I-A)^{-1}$$

where the eigenvalues of C are inside the unit circle. Solving for A yields

$$A=(C-I)(C+I)^{-1}$$

which can be substituted into (2.22) to obtain

$$(C'+I)^{-1}(C'-I)Q+Q(C-I)(C+I)^{-1}=P$$

or

$$(C'-I)Q(C+I)+(C'+I)Q(C-I)=(C'+I)P(C+I)$$

or

$$C'QC-Q=R \tag{2.26}$$

where $R=1/2(C'+I)P(C+I)$ is obviously negative definite since P is. Hence $Q-C'QC$ is positive definite. Thus, one has the matrix equation (2.26) as an alternative to Lyapunov's equation (2.22).

It must be remarked here that the solution of the Lyapunov equation (and for that matter (2.26)) may not be as simple as one tends to think. Recently, several new techniques [18–24] have been added to those available in standard monographs.

Second order differential equations

In the following chapters the systems to be studied will often be described by a set of second order differential equations, emerging from the direct application of Newton's law or Lagrangian formulation. In these cases the linearized equations of the free motion will take the general form

$$M\ddot{q}+B\dot{q}+Cq=0 \tag{2.27}$$

where the $n \times 1$ vector q represents the generalized coordinates, and M, B and C are constant n-square matrices. Normally, M will be a symmetric positive definite matrix and the systems will be classified according to the properties of the matrices B and C.

Here the connection between (2.27) and the first order differential equations (2.6) must first be established. Evidently, if A is defined as

$$A=\begin{bmatrix} -M^{-1}B & -M^{-1}C \\ I & 0 \end{bmatrix} \tag{2.28}$$

and the state vector x as

$$x=\begin{bmatrix} \dot{q} \\ q \end{bmatrix}, \tag{2.29}$$

the first order equation (2.6) and (2.27) become equivalent. It follows that the preceding discussion concerning the stability of the system (2.6) applies to systems described by (2.27). The discussion will further be expanded here.

In order to solve (2.27), one assumes solutions in the form

$$q=u\mathrm{e}^{\lambda t}; \tag{2.30}$$

substituting for q in (2.27) yields

$$(M\lambda^2 + B\lambda + C)u = 0 \tag{2.31}$$

which can be transformed into the eigenvalue problem

$$(A - \lambda I)z = 0 \tag{2.32}$$

where A is given by (2.28) and

$$z = \begin{bmatrix} \lambda u \\ u \end{bmatrix}. \tag{2.33}$$

Nontrivial solutions of (2.31) exist only if the matrix $M\lambda^2 + B\lambda + C$ is singular, that is, if λ is a root of the determinantal equation

$$\Delta(\lambda) \equiv |M\lambda^2 + B\lambda + C| = 0 \tag{2.34}$$

which is called the *characteristic equation.*

If the $2n$ roots λ_j $(j = 1, 2, \ldots, 2n)$ of (2.34) are distinct and real, one obtains $2n$ real solutions of the form (2.30) which *span* the solution space. The trivial solution $u = 0$ corresponds to the equilibrium state, and the general solution is obtained by superposition. However, note that the solution space is an n-dimensional vector space. If the matrices M, B and C are real, distinct roots occur in complex conjugate pairs $\lambda = \gamma \pm i\omega$ and the corresponding eigenvectors are also complex conjugate. The solutions satisfying (2.27) can then be expressed in the form

$$ae^{\gamma t}\cos \omega t \quad \text{and} \quad be^{\gamma t}\sin \omega t. \tag{2.35}$$

From (2.30) and (2.35) it is seen that as long as all the roots (which may also be referred to as eigenvalues or latent roots) have negative real parts, the solutions are bounded and tend to zero as $t \rightarrow \infty$. A positive real part implies an unbounded solution and hence instability.

As discussed earlier, coincident eigenvalues (multiple roots) may be associated with solutions which contain powers of t multiplied by the exponential functions. *Stability properties, however, do not change provided the roots do not have zero real parts.* The *critical* case in which some roots have zero real parts, while the remaining roots have negative real parts, may be stable or unstable.

On the basis of the foregoing discussion and the roots

$$\lambda_j = \gamma_j + i\omega_j; \qquad j = 1, 2, \ldots, 2n, \tag{2.36}$$

where γ_j and ω_j are real, the following three cases will be identified with regard to the stability of the equilibrium state $q=\dot{q}=0$:

(a) All $\gamma_j<0$: asymptotic stability.

(b) At least one $\gamma_j>0$: instability.

(c) If some $\gamma=0$ while the remaining γ's are negative the state is, in general, critical, particularly if the system is a linearized model of a nonlinear system. Asymptotic stability is now ruled out but not stability. Indeed, the state is stable or unstable according to whether or not the *multiplicity* of λ_j (with $\gamma_j=0$) is equal to the *degeneracy* of the λ-matrix $M\lambda^2+B\lambda+C$ when evaluated at $\lambda=\lambda_j$; i.e., the number of linearly independent eigenvectors corresponding to λ_j must be equal to its multiplicity for stability.

The following special cases are of particular interest:

(i) If all $\gamma_j=0$ and ω is not repeated, or more generally, if the multiplicity of the repeated roots is equal to the degeneracy of the λ-matrix, the state is obviously stable, and this case is not even referred to as *critical* in the literature concerning *elastic stability.* A truly critical situation arises when at least a pair of ω's vanishes as well (assuming complex conjugate roots), such that a decision on stability or instability requires higher order terms. It is of course understood that the linear system here is viewed as a linearized model of a nonlinear system; otherwise, the linear system is unstable because terms with t appear in the solution. This case will arise in Chapter 3.

(ii) If all $\gamma_j=0$ and at least a pair of ω's coincide such that the multiplicity exceeds the degeneracy, then the linear system is again unstable since terms with t appear in the solution. In this case the number of linearly independent eigenvectors corresponding to the root ω with multiplicity k is less than k, and a λ-matrix is called *defective* if the *multiplicity* of a root exceeds the dimension of its eigenvector space. This case will arise in Chapters 5 and 6.

In the following chapters two types of instability will be identified, and they are defined here as follows:

(I) *Divergence*: This instability is identified by

$$\gamma_j>0, \qquad \omega_j=o. \tag{2.37}$$

(II) *Flutter*: This instability is identified by

$$\gamma_j>0, \qquad \omega_j\neq 0. \tag{2.38}$$

Similarly, the critical state $\gamma_j = 0$ is also called divergence or flutter according to the above definitions (e.g., cases (i) and (ii) respectively). Divergence instability is a statical one and is associated with a branching of the solution (i.e., bifurcation of two equilibrium paths, considering the behaviour when a parameter η is varied). Flutter, on the other hand, is a dynamic loss of stability characterized by oscillations with increasing amplitude. An interesting type of flutter instability occurs in conjunction with the stability criterion (c,ii). Assuming again distinct complex conjugate imaginary roots and considering the behaviour as a function of a parameter η, at one stage two pairs of roots may coincide and, upon a further increase in η, may become complex conjugate, inducing flutter instability. The onset of flutter is characterized by the coincidence of imaginary roots. This sort of behaviour is demonstrated by undamped circulatory systems and will be discussed in Section (2.5), as well as in Chapter 6. It must be noted again, however, that the coincidence of imaginary roots is necessary but not sufficient for flutter to occur; the eigenvectors corresponding to repeated roots must also coincide, effectively reducing the dimension of the space spanned by the eigenvectors.

The Routh-Hurwitz criterion

The characteristic equation (2.34) yields a characteristic polynomial which can be expressed as

$$f(\lambda) = \lambda^m + a_1\lambda^{m-1} + a_2\lambda^{m-2} + \cdots + a_m = 0, \qquad m = 2n \tag{2.39}$$

since the matrix M is assumed to be positive definite. The coefficients a_1 $(i = 1, 2, \ldots, m)$ are also real, because B and C are assumed to be real. The question then arises whether one can derive information about the signs of the roots on the basis of the coefficients a_i without actually finding the roots. This question has been dealt with by Routh and Hurwitz independently.

The polynomial (2.39) is called a *Hurwitz polynomial* if all its roots have negative real parts. The following criterion can then be stated:

The necessary and sufficient condition for the polynomial $f(\lambda)$ to be a Hurwitz polynomial is that all the principal minors of the *Hurwitz matrix*

$$H = \begin{bmatrix} a_1 & 1 & 0 & 0 & 0 & \cdots & 0 \\ a_3 & a_2 & a_1 & 1 & 0 & \cdots & 0 \\ a_5 & a_4 & a_3 & a_2 & a_1 & 1 \cdot & \cdot \\ \cdots & \cdots & \cdots & \cdots & \cdots & \cdots & a_m \end{bmatrix} \tag{2.40}$$

are positive. That is,

$$\Delta_1 = a_1 > 0,$$

$$\Delta_2 = \begin{vmatrix} a_1 & 1 \\ a_3 & a_2 \end{vmatrix} > 0, \tag{2.41}$$

$$\cdots\cdots\cdots\cdots\cdots$$

$$\Delta_m = a_m \Delta_{m-1} > 0.$$

Note that the Hurwitz matrix is m-square and it is constructed such that all $a_k = 0$ for $k > m$.

The proof [25, 15] can be based upon the properties of the Hurwitz matrices associated with a Hurwitz polynomial $f(\lambda)$ of degree m and a Hurwitz polynomial $\bar{f}(\lambda)$ of degree $m+1$. Let

$$\bar{f}(\lambda) = (\lambda + 2a)\, f(\lambda) + \lambda f^*(\lambda)$$

where $a > 0$ and

$$f^*(\lambda) = (-1)^m f(-\lambda).$$

It follows that

$$\bar{f}(\lambda) = 2[\lambda^{m+1} + a\lambda^m + (aa_1 + a_2)\lambda^{m-1} + aa_2\lambda^{m-2} + \cdots].$$

It is not difficult to show [25, 15] that if $f(\lambda)$ is a Hurwitz polynomial, so is $\bar{f}(\lambda)$. The Hurwitz matrix of $\frac{1}{2}\bar{f}(\lambda)$ can then be written as

$$\bar{H} = \begin{bmatrix} a & 1 & 0 & 0 & \cdots\cdots \\ aa_2 & aa_1 + a_2 & a & 1 & \cdots\cdots \\ aa_4 & aa_3 + a_4 & aa_2 & aa_1 + a_2 & \cdots\cdots \\ \cdots & \cdots & \cdots & \cdots & \cdots\cdots \end{bmatrix}$$

with the principal minors

$$\bar{\Delta}_1 = a,$$

$$\bar{\Delta}_2 = a^2 \Delta_1,$$

$$\bar{\Delta}_3 = a^3 \Delta_2,$$

$$\cdots\cdots\cdots\cdots$$

$$\bar{\Delta}_{m+1} = a^{m+1} \Delta_m.$$

Clearly, if $\Delta_i > 0$ for $i = 1, 2, \ldots, m$; then $\bar{\Delta}_i > 0$ for $i = 1, 2, \ldots, m+1$ since $a > 0$. On the other hand, if all $\bar{\Delta}_i > 0$, then all $\Delta_i > 0$. It follows that, if the Hurwitz criterion holds for polynomials of degree m, a Hurwitz polynomial of degree $m+1$ can be formed from that of degree m. Since the Hurwitz criterion holds for polynomials of degree one and two (as can easily be verified), by induction, it also holds for polynomials of degree m.

2.4 Classification of systems

It was remarked in the preceding section that the second order differential equation (2.27) may emerge as a result of the so-called Lagrangian formulation. Assuming again that the state or position of a dynamical system with n degrees of freedom can be specified completely by n generalized coordinates $q_i (i = 1, 2, \ldots, n)$, the concepts of *kinetic energy* and *work* done by the force system involved can be used as a basis to explore the dynamic properties of the system. The total kinetic energy T of the system may, in general, depend on the q_i as well as their time derivatives $\dot{q}_i$. The work done by the system of forces when the q_i are given small variations δq_i, on the other hand, can be expressed as

$$\langle F, \delta q \rangle = F_1\, \delta q_1 + F_2\, \delta q_2 + \cdots + F_n\, \delta q_n \tag{2.42}$$

where the F_i are *generalized forces*, one reason for this terminology being that the F_i may not have the dimension of a true force. If q_i is an angular displacement, for example, the corresponding F_i is a moment. Also, the q_i may represent currents, electric charges, areas, volumes, etc., depending on applications in which $F_i\, \delta q_i$ has the dimension of *work*.

Lagrange's equations can be derived from Hamilton's principle which will now be discussed briefly.

Hamilton's Principle

The fundamental equations governing a variety of physical phenomena can be derived from a basic principle due to Hamilton. To facilitate the discussion, the dynamics of a single particle will be considered first, but the extension to a system of particles follows immediately, and to other situations by analogy.

Let the particle of mass m move in a force field; then

$$m\ddot{r} = f \tag{2.43}$$

where r is the position vector measured from the origin of a fixed reference frame and the vector f represents the force acting on the particle. Consider a variation δr of the actual path between $t = t_1$ and $t = t_2$, at which instances $\delta r = 0$ and the actual and virtual paths coincide. Here, t_1 and t_2 are arbitrary. The scalar product of δr with (2.43) yields, upon integration,

$$\int_{t_1}^{t_2} m\,\ddot{r}\cdot\delta r\,dt=\int_{t_1}^{t_2} f\cdot\delta r\,dt.$$

Integrating the left hand side by parts results in

$$m[\dot{r}\cdot\delta r]_{t_1}^{t_2}-m\int_{t_1}^{t_2}\dot{r}\cdot\delta\dot{r}\,\mathrm{d}t=\int_{t_1}^{t_2} f\cdot\delta r\,\mathrm{d}t.$$

Since

$$\dot{r}\cdot\delta\dot{r}=\tfrac{1}{2}\,\delta\dot{r}^2$$

and $\delta r=0$ at $t=t_1$ and $t=t_2$, one has

$$-\tfrac{1}{2}m\int_{t_1}^{t_2}\delta\dot{r}^2\,\mathrm{d}t=\int_{t_1}^{t_2} f\cdot\delta r\,\mathrm{d}t.$$

But $\delta(\frac{1}{2}m\,\dot{r}^2)$ is immediately recognized as the variation of the kinetic energy, δT, and hence

$$\int_{t_1}^{t_2}(\delta T+f\cdot\delta r)\,\mathrm{d}t=0 \tag{2.44}$$

which is the most general form of Hamilton's principle for a particle.

The kinetic energy of a system of n particles can be expressed as

$$T=\tfrac{1}{2}m_i(\dot{r}^i)^2,$$

and the total work done by the forces acting on the particles m_i is

$$f^i\cdot\delta r^i$$

where the summations over i range from 1 to n. Hamilton's principle, as expressed by (2.44), applies to this system of particles as well as other dynamical systems described by n generalized coordinates. In the case of continuous systems, integrations replace summations.

Suppose now that the forces acting on the system are derivable from a *potential function* V such that

$$f^i\cdot\delta r^i=-\delta V;$$

Hamilton's principle then takes the simple form

$$\delta\int_{t_1}^{t_2}(T-V)\,\mathrm{d}t=0. \tag{2.45}$$

Forces which can be derived from a potential function are *conservative*, and if the forces acting on a system are conservative, then, the system is said to be *conservative.* In mechanics the potential function V is the *potential energy* and the difference between the kinetic and potential energies,

$$L = T - V,$$

is often called the *Lagrangian* function. The variational problem characterized by Hamilton's principle can then be expressed as follows: *The motion of a conservative system is such that the integral of the Lagrangian between arbitrary points in time is stationary for the actual motion.*

If the forces acting on a system cannot be derived from a potential they are *non-conservative.* The work done by non-conservative forces during a displacement from A to B depends on the path followed in the motion from A to B and not just on these initial and final points. Follower forces and frictional forces are examples of non-conservative forces. The nature of forces will be taken up again when the classification of systems is discussed in the sequel.

Lagrange's equations

It will be assumed throughout that the systems described by the generalized coordinates q_i $(i = 1, 2, \ldots, n)$ are *scleronomic* and *holonomic.* The first property dictates that the position vector of an arbitrary material element of the system can be expressed in terms of the q_i only and is not a function of t explicitly; i.e., $r = r[q_i(t)]$. On the basis of the second property the δq_i represent an admissible displacement, and each q_i can be varied arbitrarily and independently. It follows that

$$\delta r = \frac{\partial r}{\partial q_i}\, \delta q_i,$$

and hence the work done by the force system in small displacements δq_i can be expressed as

$$f^i \cdot \delta r^i = F_1\, \delta q_1 + F_2\, \delta q_2 + \cdots + F_n\, \delta q_n = \langle F, \delta q\rangle.$$

In general, the generalized forces F_i are not derivable from a potential. If the system of forces is conservative, however, one has

$$F_1\, \delta q_1 + F_2\, \delta q_2 + \cdots + F_n\, \delta q_n = -\delta V \tag{2.46}$$

and

$$F_i = -\frac{\partial V}{\partial q_1}, \qquad i = 1, 2, \ldots, n.$$

Lagrange's equations follow directly from Hamilton's principle which now takes the form

$$\delta\int_{t_1}^{t_2} T\,\mathrm{d}t + \int_{t_1}^{t_2} F_i\,\delta q_i\,\mathrm{d}t = 0. \tag{2.47}$$

Since the kinetic energy is, in general, a function of both the q_i and $\dot{q}_i$,

$$\begin{aligned}\delta T &= \frac{\partial T}{\partial q_i}\,\delta q_i + \frac{\partial T}{\partial \dot{q}_i}\,\delta\dot{q}_i\\ &= \frac{\partial T}{\partial q_i}\,\delta q_i + \frac{\partial T}{\partial \dot{q}_i}\frac{\mathrm{d}}{\mathrm{d}t}\,\delta q_i\\ &= \frac{\partial T}{\partial q_i}\,\delta q_i + \frac{\mathrm{d}}{\mathrm{d}t}\left(\frac{\partial T}{\partial \dot{q}_i}\,\delta q_i\right) - \frac{\mathrm{d}}{\mathrm{d}t}\left(\frac{\partial T}{\partial \dot{q}_i}\right)\delta q_i.\end{aligned}$$

Using this result in (2.47) yields

$$\int_{t_1}^{t_2}\left[\frac{\partial T}{\partial q_i} - \frac{\mathrm{d}}{\mathrm{d}t}\frac{\partial T}{\partial \dot{q}_i} + F_i\right]\delta q_i\,\mathrm{d}t + \left[\frac{\partial T}{\partial \dot{q}_i}\,\delta q_i\right]_{t_1}^{t_2} = 0.$$

But $\delta q_i = 0$ at $t = t_1$ and $t = t_2$, and the coefficients of the independent variations must, therefore, vanish, leading to

$$\frac{\mathrm{d}}{\mathrm{d}t}\frac{\partial T}{\partial \dot{q}_i} - \frac{\partial T}{\partial q_i} = F_i \qquad i = 1, 2, \ldots, n. \tag{2.48}$$

These equations are known as Lagrange's equations and are valid whether the forces involved are conservative or not. In the special case of conservative forces, $F_i = -(\partial V/\partial q_i)$, and (2.48) can be written as

$$\frac{\mathrm{d}}{\mathrm{d}t}\frac{\partial T}{\partial \dot{q}_i} - \frac{\partial T}{\partial q_i} + \frac{\partial V}{\partial q_i} = 0 \tag{2.49}$$

or

$$\frac{\mathrm{d}}{\mathrm{d}t}\frac{\partial T}{\partial \dot{q}_i} - \frac{\partial L}{\partial q_i} = 0 \tag{2.50}$$

where L is the *Lagrangian*.

Clearly, the potential energy V is a function only of position and represents the total work of internal and external forces; it does not depend on the generalized velocities. Expanding the potential energy into Taylor series in the vicinity of the equilibrium state $q_i = 0$ yields [10]

$$V = v_0 + k_i q_i + \frac{1}{2!}\,k_{ij}q_iq_j + \frac{1}{3!}\,k_{ijl}q_iq_jq_l + \cdots$$

where k_i, k_{ij}, etc. are derivatives of V with respect to the q_i evaluated at the equilibrium point. v_0 is the potential energy at the equilibrium point and can be assumed zero without loss of generality. The derivatives k_i, on the other hand, are necessarily zero, because the point under consideration is one of equilibrium and the principle of virtual work applies. It follows that a first order approximation for the potential energy is in the form

$$V(q_i)=\tfrac{1}{2}k_{ij}q_iq_j=\tfrac{1}{2}\langle q, Kq\rangle \tag{2.51}$$

where the matrix K is real symmetric.

The kinetic energy T can also be expressed as a quadratic form (in $\dot{q}_i$) by making use of

$$\dot{r}=\frac{\partial r}{\partial q_j}\frac{\mathrm{d}q_j}{\mathrm{d}t}.$$

Thus,

$$T=\tfrac{1}{2}m_{ij}\dot{q}_i\dot{q}_j=\tfrac{1}{2}\langle \dot{q}, M\dot{q}\rangle \tag{2.52}$$

where the coefficients m_{ij} are, in general, functions of the q_i. If the attention is confined to small displacements, the Taylor series expansion of the coefficients m_{ij} (q_k) can be used in (2.52) to obtain a quadratic form in the $\dot{q}_i$ with constant coefficients as a first approximation [consistent with (2.51)] for T. In this book, M will always be taken as a constant real symmetric matrix. Furthermore, it is clear from its definition that the kinetic energy is always positive, and the quadratic form (2.52) is, therefore, positive definite. It follows that the matrix M is positive definite.

Using (2.51) and (2.52) in Lagrange's equations (2.49) yields the second order matrix equation

$$M\ddot{q}+Kq=0 \tag{2.53}$$

for a conservative system.

As remarked earlier, forces acting on a system are not always conservative and cannot always be derived from a potential. Such forces may be classified into two main categories for the purposes of this book:

(1) velocity independent: $F(q_i)$
(2) velocity dependent: $F(q_i, \dot{q}_i)$.

While the first class obviously contains conservative forces, a third class, namely, instationary forces of the type $F(q_i, \dot{q}_i, t)$ is not even mentioned here. Evidently, the attention is now to be focussed on forces which are not conservative but do not depend on time t explicitly. In the first category, one has *follower* type forces which will be classified as *circulatory* [13]. They arise in contemporary mechanical, aeronautical and missile engineering. Mathematically, linear generalized forces which can be expressed as

$$F_i = -k_{ij}q_j \qquad i, j = 1, 2, \ldots, n \tag{2.54}$$

where k_{ij} is a constant but asymmetric matrix, are associated with circulatory forces acting on the system. Since an asymmetric matrix can always be expressed as the sum of a symmetric and skew-symmetric matrix, it is clear that circulatory forces are characterized by the skew-symmetric component of k_{ij}. Lagrange's equations now yield

$$M\ddot{q} + Kq = 0 \tag{2.55}$$

where K is asymmetric. Systems described by (2.55) will be called *circulatory* when the matrix K is strongly asymmetric such that the matrix

$$M^{-1}K \tag{2.56}$$

is not *symmetrizable* (Chapter 1).

If it turns out that $M^{-1}K$ is a symmetrizable matrix, the systems described by (2.55) will be called *pseudo-conservative*. It will be seen that the behaviour of circulatory and pseudo-conservative systems is quite distinct, the latter exhibiting features analogous to those of conservative systems. If K is symmetric the system (2.55) is of course conservative.

Three types of velocity-dependent forces will be considered. Linear generalized forces, given by

$$F_i = -g_{ij}\dot{q}_j \qquad i, j = 1, 2, \ldots, n \tag{2.57}$$

where g_{ij} is a constant skew-symmetric matrix, are associated with *gyroscopic forces* (e.g., Coriolis forces, Lorentz forces). The differential equations of a gyroscopic system, obtained from Lagrange's equations, are in the form

$$M\ddot{q} + G\dot{q} + Kq = 0 \tag{2.58}$$

where $G = -G'$. If $K = K'$, then (2.58) describes a *gyroscopic conservative* system which will be discussed in detail in Chapter 5. It is noted here, however, that the work done by gyroscopic forces in a real motion is zero. This can easily be seen by considering the total work performed by the generalized forces F_i per unit time which can be expressed as

$$\langle F, \dot{q}\rangle = \langle -G\dot{q}, \dot{q}\rangle = 0.$$

In other words, a force of this type is perpendicular to the velocity vector at the point upon which it acts. It must immediately be emphasized, however, that the work done by gyroscopic forces in a virtual displacement is rarely zero [13].

An important class of velocity-dependent forces is related to dissipation. The most common dissipative forces (e.g., viscous damping) can be described by the generalized forces

$$F_i = -\mathrm{d}_{ij}\dot{q}_j \qquad i, j = 1, 2, \ldots, n \tag{2.59}$$

where the matrix D is symmetric and non-negative definite. The F_i can alternatively be considered as derivable from the so-called Rayleigh dissipation function

$$R = \tfrac{1}{2}\langle \dot{q}, D\dot{q}\rangle \tag{2.60}$$

as

$$F_i = -\frac{\partial R}{\partial \dot{q}_i} \tag{2.61}$$

Physically, R cannot be negative and the matrix D, therefore, is positive definite or positive semi-definite.

The differential equations of a gyroscopic damped system can then be obtained from Lagrange's equations as

$$M\ddot{q} + (D + G)\dot{q} + Kq = 0. \tag{2.62}$$

Clearly, the undamped nongyroscopic system ($D = G = 0$) here is conservative or non-conservative depending on whether K is symmetric or asymmetric.

Finally, a third type of velocity-dependent force will be introduced. It has been observed in a variety of applications concerning gyroscopic systems that dissipative forces derivable from a dissipation function of the form

$$\bar{R} = \tfrac{1}{2}\langle \dot{q}, D_e\dot{q}\rangle + \langle \dot{q}, Sq\rangle \tag{2.63}$$

may act on the system. Here, the matrix D_e is symmetric non-negative definite and S is skew-symmetric. The generalized force vector is given by

$$F = -D_e\dot{q} - Sq, \tag{2.64}$$

and this type of dissipation is sometimes called "*constraint damping*" in the study of dissipative gravity oriented sattelites. *External damping* associated with rotating shafts is of a similar nature and will be discussed extensively in Chapter 5.

Taking into account all the forces discussed so far, Lagrange's equations yield

$$M\ddot{q} + (D + D_e + G)\dot{q} + (K + S)q = 0 \tag{2.65}$$

or

$$M\ddot{q}+B\dot{q}+Cq=0 \tag{2.66}$$

where $C=K+S$ and $B=D+D_e+G$ are general $n\times n$ matrices.

The following four chapters will deal with

1. conservative systems,
2. pseudo-conservative systems,
3. gyroscopic systems, and
4. circulatory systems,

the effect of dissipative forces being considered in each case separately. All systems considered are autonomous and holonomic.

2.5 Multiple-parameter systems and associated definitions

As emphasized earlier, one of the main goals of this book is to present a multiple-parameter theory. If the behaviour of a system is influenced by several independently assignable parameters, it is natural to inquire how the system will behave as these parameters are varied gradually. In mechanics, for example, one may have several independent loads and the effect of these loads on the instability behaviour of an initially stable system is of vital importance in a design process.

In order to trace the effect of such parameters, the differential equations discussed in the preceding section must be expressed in a more explicit form, incorporating the parameters whose influence on the dynamic response and instability behaviour of the system is sought.

It is convenient to consider first the conservative systems described by (2.53). The potential energy V can normally be expressed as the sum of the internal energy (e.g., strain energy) U, which is a function of generalized coordinates only, and the potential energy associated with external parameters η_k $(k=1,2,\ldots,m)$. Thus, it will be assumed that

$$V=U(q_i)-\eta_k E^k(q_i) \qquad k=1,2,\ldots,m \tag{2.67}$$

where $\eta_k E^k(q_i)$ is the total work done by the η_k over the corresponding displacements $E^k(q_i)$. In view of (2.51), the quadratic approximation of (2.67) is in the form

$$\begin{aligned} V&=\tfrac{1}{2}(u_{ij}-\eta_k e^k_{ij})q_iq_j \\ &=\tfrac{1}{2}\langle q,(U-\eta_k E^k)q\rangle \end{aligned} \tag{2.68}$$

where the matrix U will always be symmetric positive definite, assuming that the state $\eta_k = 0$ is stable. The matrices E^k are symmetric but not necessarily positive definite.

Introducing (2.68) into Lagrange's equations yields

$$M\ddot{q} + (U - \eta_k E^k)q = 0 \tag{2.69}$$

which, upon setting $q = ue^{\lambda t}$, results in the characteristic equation

$$\Delta(\lambda, \eta_k) \equiv |M\lambda^2 + U - \eta_k E^k| = 0. \tag{2.70}$$

Considering now circulatory systems, the η_k represent circulatory forces, and although internal energy remains in the same form, the matrices E^k are no longer symmetric and one has a characteristic equation of the form (2.70) with asymmetric matrices E^k.

In the case of gyroscopic systems, at least one parameter will also appear in the matrix G as well as K in equation (2.58), leading to a characteristic equation which is again a function of η_k as well as λ.

Generally, the eigenvalues λ_i are complex; however, in a number of important systems, and in particular when the dissipative forces are not present, the characteristic equation is a function of λ^2 (rather than λ) which yields real roots λ_i^2 at least for some values of η_k. In the case of conservative systems (2.70), for example, λ_i^2 are always real and at $\eta_k = 0$ they supply the *natural frequencies;* as η_k are varied, at each specific set of values given to η_k, *the corresponding frequencies of free vibration* are obtained. Thus, one may consider

$$\Delta(\lambda^2, \eta_k) = 0 \tag{2.71}$$

as the equation of certain m dimensional surfaces in the $m+l$ dimensional $\lambda^2 - \eta_k$ space which are called *characteristic surfaces.* In some situations it may be convenient to replace λ^2 by ω^2, which are related by $\lambda = i\omega$ $(i = \sqrt{-1})$.

Normally, $\eta_k = 0$ represents a stable state and attention is then focussed on the critical values of η_k at which the system loses its stability. Considering an arbitrary path emanating from the origin of the parameter space, the initially stable system may lose stability by divergence or by flutter according to whether one root λ^2 vanishes or two roots (and corresponding to coincident eigenvalues (with coincident eigenvectors) the two roots become complex conjugate upon passing through the critical point.

The locus of the points in the η_k space corresponding to zero roots (onset of divergence) is called the *divergence boundary*, while the points corresponding to coincident eigenvalues (with coincident eigenvectors) comprise the *flutter boundary*. The points on these two boundaries which are associated with an *initial* loss of stability form the *stability boundary*.

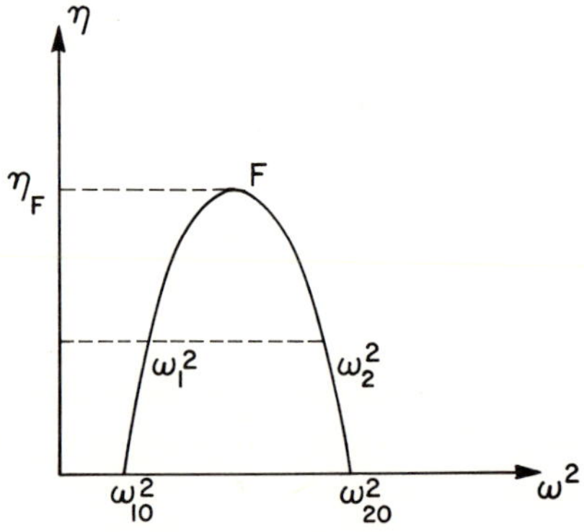

Figure 2.3

Note that the definitions of divergence and flutter boundaries allow for instabilities associated with higher modes and do not necessarily imply an initial loss of stability. It follows that setting $\lambda^2=0$ in (2.71) yields the divergence boundary which may consist of several surfaces. Similarly, the flutter boundary is the locus of coincident eigenvalues ω^2 which are real on one side of the boundary and complex conjugate on the other.

Considering a single parameter η only, a typical *characteristic curve* is sketched in Figure 2.3 to illustrate how flutter occurs. It is seen that for $\eta=0$ there are distinct real roots on the ω^2 axis which yield the natural frequencies of the system. As η is increased, the two distinct roots approach each other, and eventually coalesce at the point *F*, the onset of flutter. Upon a further increase in η, they become complex conjugate, resulting in a root with a positive real part which implies self-excited oscillations with increasing amplitude. It is clear from Figure 2.3 that *F* is an extremum point on the characteristic curve and it should, therefore, satisfy

$$\frac{d\eta}{d\omega^2}=0 \tag{2.72}$$

which is equivalent to

$$\frac{\partial\Delta}{\partial\omega^2}=0 \tag{2.73}$$

since

$$\frac{\partial\Delta}{\partial\omega^2}+\frac{\partial\Delta}{\partial\eta}\frac{d\eta}{d\omega^2}=0.$$

Equation (2.73) also follows from the double-root condition, assuming of course continuity for Δ. In the presence of multiple parameters, one notes that $\Delta(\omega^2, \eta_k)=0$ is associated with m independent variables and choosing ω^2 as one of these variables yields

$$\frac{\partial\Delta}{\partial\omega^2}+\frac{\partial\Delta}{\partial\eta_k}\frac{\partial\eta_k}{\partial\omega^2}=0 \qquad \begin{array}{l} k=1,2,\ldots,m \\ \text{(no summation)} \end{array}$$

where each η_k is given the role of a dependent variable in turn. At the flutter point all $\partial\eta_k/\partial\omega^2=0$ and hence (2.73) holds for multiple-parameter systems as well. Alternatively, if (2.73) is accepted on the basis of a double-root, then $\partial\eta_k/\partial\omega^2=0$ follows.

The concepts of flutter boundary, divergence boundary, etc. can be extended to dissipative systems which are normally associated with complex roots even when $\eta_k=0$. In these cases, the complex roots λ of the initially stable system have negative real parts; as η_k are varied, stability may be lost either by divergence or by flutter. Flutter occurs when the real part of a root passes through zero and becomes positive, and the locus of all such points defines the flutter boundary. Divergence is again associated with vanishing roots, and the definitions of divergence and stability boundaries remain the same as before.

In the following chapters, various classes of systems will be studied in detail and additional stability criteria relevant to the class under consideration will be discussed. Properties of characteristic surfaces and of divergence, flutter and stability boundaries will be explored and certain illustrative examples will be analyzed.

3

Conservative systems

Conservative systems were defined in Section 2.4, and the stability of linear conservative systems will now be studied in some detail, including the effect of damping on the stability properties. It is understood that although the addition of gyroscopic forces does not render the system nonconservative, systems associated with gyroscopic forces will be discussed separately in Chapter 5, and the term "conservative systems" always refers to nongyroscopic conservative systems here.

3.1 Stability of conservative systems

The first approximations for the kinetic and potential energies of a conservative system were found (Section 2.4) to be of the form

$$T(\dot{q}) = \tfrac{1}{2}\langle \dot{q}, M\dot{q}\rangle = \tfrac{1}{2}m_{ij}\dot{q}_i\dot{q}_j$$

and (3.1)

$$V(q) = \tfrac{1}{2}\langle q, Kq\rangle = \tfrac{1}{2}k_{ij}q_iq_j, \qquad i, j = 1, 2, \ldots, n$$

where $M = M'$, $K = K'$ and $M > 0$. Making use of Lagrange's equations yields

$$M\ddot{q} + Kq = 0 \tag{3.2}$$

which governs the motion in the vicinity of the equilibrium state $q = \dot{q} = 0$.

Defining the *Hamiltonian* as

$$H = T + V, \tag{3.3}$$

one observes that H represents the total energy associated with conservative systems. If V is positive definite, clearly H is also positive definite. Since the well-known law of conservation of energy requires that the total energy must be conserved during the motion, $\dot{H} = 0$. It follows that when $V > 0$, the Hamiltonian H is a Lyapunov function, and the state $q = \dot{q} = 0$ is stable. Thus, Lagrange's theorem is established: *An equilibrium state is stable if the total potential energy has a complete relative minimum at that point.*

Note that a complete relative minimum of the potential energy V is ensured when $K>0$; however, if $K\geq 0$ (positive semi-definite), a complete relative minimum may still exist depending on the higher order terms of V. A full discussion concerning the stability of such critical points based on a nonlinear analysis can be found in Ref. [10]. It is further noted that Lagrange's theorem supplies only a sufficient condition for stability; whether a complete relative minimum is also a necessary condition apparently has not yet been solved in complete generality. Koiter [26] has tackled this problem by introducing a positive definite damping into the system, and showed that an equilibrium state in which the potential energy has no proper relative minimum is always unstable in the presence of damping with a positive definite energy dissipation. Recently, Hagedorn [27] demonstrated that if the potential energy has continuous second derivatives and it assumes a strong relative maximum at an equilibrium state, then that state is unstable.

Consider now the eigenvalue problem associated with (3.2). Assuming solutions of the form

$$q = u\mathrm{e}^{i\omega t} \quad \text{where} \quad i=\sqrt{-1} \tag{3.4}$$

leads to

$$(-\omega^2 M + K)u = 0, \tag{3.5}$$

and since M is positive definite and K symmetric, it follows from Sections 1.4 and 1.5 that the matrix pencil $-\omega^2 M+K$ is *simple*, and the matrix $M^{-1}K$ is *symmetrizable*. Hence all the eigenvalues ω^2 associated with (3.5) are real, and one deduces from (3.5) that the eigenvectors are also real. An eigenvalue ω^2 can be expressed in terms of the corresponding eigenvector u as

$$\omega^2 = \frac{\langle u, Ku\rangle}{\langle u, Mu\rangle}. \tag{3.6}$$

Since the matrix pencil $-\omega^2 M+K$ is simple, corresponding to the frequencies ω_j^2 $(j=1,2,\ldots,n)$ there exist n *linearly independent* eigenvectors u^j (whether ω_j^2 are distinct or not), and the solutions

$$\left.\begin{aligned} q^j &= u^j \mathrm{e}^{i\omega_j t} \\ q^j &= u^j \mathrm{e}^{-i\omega_j t} \end{aligned}\right\} \quad j=1,2,\ldots,n \tag{3.7}$$

span the solution space. The general solution can be expressed as a linear combination of (3.7).

An important distinction must now be noted between treating $-\omega^2 M+K$ as a matrix pencil in ω^2 and as an ω-matrix in ω. As emphasized before, in the former case the pencil in ω^2 is *simple* and has a full complement of linearly independent eigenvectors even when the eigenvalues ω^2 are repeated. If, for example, $\omega_1^2=\omega_2^2\neq 0$, the corresponding vectors u^1 and u^2 can still be considered linearly independent and the associated solutions are of the form

$$u^1 e^{i\omega_1 t}, \quad u^1 e^{-i\omega_1 t}, \quad u^2 e^{i\omega_1 t}, \quad u^2 e^{-i\omega_1 t}. \tag{3.8}$$

Thus, it is observed that the coincidence of eigenvalues ω^2 does not result in solutions which are multiplied by t as proposed wrongly by Lagrange [28]. Furthermore, both the matrix pencil in ω^2 and the ω-matrix are *simple* because there exist linearly independent eigenvectors associated with repeated roots. It is, however, another matter if one of the eigenvalues vanishes. Suppose $\omega_1^2=0$; clearly, the pencil in ω^2 remains simple but not the ω-matrix which now has zero as a repeated root and u^1 as the only corresponding eigenvector. In other words, the multiplicity of ω_1 now exceeds the number of corresponding linearly independent eigenvectors, and the ω*-matrix* is *defective.* The associated solutions are of the form

$$u^1, \quad u^1 t,$$

and the general solution, therefore, contains t explicitly.

Alternatively, if the second order differential equation (3.2) is transformed into a first order equation, the eigenvalue problem

$$(A-\omega I)z=0, \tag{3.9}$$

where

$$A=\begin{bmatrix} 0 & M^{-1}K \\ I & 0 \end{bmatrix} \quad \text{and} \quad z=\begin{bmatrix} \omega u \\ u \end{bmatrix},$$

yields the roots ω of (3.5), and corresponding to the repeated root $\omega_1=0$ one has the single eigenvector $(0, u^1)$. Clearly, the *multiplicity* of the eigenvalue $\omega_1=0$ exceeds its *index* and the matrix A is no longer of simple structure, resulting in solutions which contain t as can be deduced from (2.20).

On the basis of the foregoing discussion, it may be concluded that as long as the eigenvalues ω^2 of (3.5) are positive, the solutions (3.7) remain bounded and the equilibrium state $q=\dot{q}=0$ is stable. Since all $\omega^2>0$ if K is positive definite, it is observed as before that the system is stable if the quadratic form of its potential energy is positive definite. At the critical point when one of the eigenvalues ω^2 vanishes the linear system is unstable, because the solutions (3.7) contain t explicitly. In this case, however, the quadratic form of the potential energy is positive semi-definite and a complete relative minimum for V is not ruled out. Indeed, a nonlinear analysis involving higher order terms is required to ascertain whether or not the potential energy V has a minimum at the origin. If a minimum exists, then a nonlinear dynamical analysis will also show that the critical state is stable. Such an analysis will not be pursued here.

Finally, it is observed that if any of the eigenvalues ω^2 is negative, the general solution contains unbounded terms and the equilibrium state is unstable. This happens when the quadratic form $\langle q, Kq\rangle$ becomes indefinite, admitting negative values.

Before proceeding to the next section, note that the case in which $\omega_1^2 = \omega_2^2 = 0$ can be handled similarly. While the matrix pencil in ω^2 in this case remains simple with distinct eigenvectors u^1 and u^2 corresponding to the repeated root $\omega_1^2 = 0$, the ω-matrix has $\omega_1 = 0$ as a root repeated four times with only two distinct eigenvectors, and the solutions are, therefore, of the form

$$u^1, \quad u^1 t, \quad u^2, \quad u^2 t.$$

Alternatively, $\omega_1 = 0$ can be considered as a 4-fold eigenvalue of the matrix (3.9) whose *index* is two, the eigenvectors $(0, u^1)'$ and $(0, u^2)'$ also being repeated. From either point of view, one observes that the general solution contains t explicitly.

3.2 Convexity of the fundamental characteristic surface

Suppose now that the system is under the influence of several independent parameters $\eta_k (k = 1, 2, \ldots, m)$ such that the potential energy in (3.1) takes the explicit form

$$V = \tfrac{1}{2}\langle q, (u - \eta_k E^k) q\rangle \tag{3.10}$$

as discussed Section 2.5. Here, U is associated with internal energy and will always be assumed to be positive definite while the matrices E^k $(k = 1, 2, \ldots, m)$ are symmetric but not necessarily positive definitite.

The corresponding characteristic equation

$$\Delta(\omega^2, \eta_k) \equiv |-\omega^2 M + U - \eta_k E^k| = 0 \tag{3.11}$$

defines certain m dimensional *characteristic surfaces* in the $m+1$ dimensional $\omega^2 - \eta_k$ space and the part of this surface which consists of points closest to the origin is called the *fundamental characteristic surface* (Fig. 3.1). The region containing the origin of the $\omega^2 - \eta_k$ space and bounded

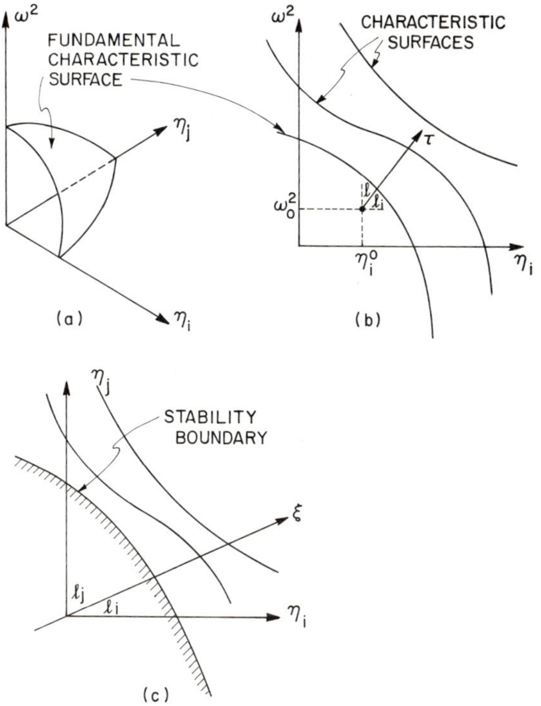

Figure 3.1

by the fundamental characteristic surface, which incidentally may be composed of more than one self-curvilinear surface, is called the *fundamental region.* Clearly, the *stability boundary* is associated with the fundamental characteristic surface and can be obtained as the intersection of the latter surface with the plane $\omega^2 = 0$. The intersections of the characteristic surfaces (3.11) with $\omega^2 = 0$ will, in general, be called *divergence boundary.* It is understood that while the stability boundary is associated with an initial loss of stability, the divergence boundary consists of points where some eigenvalue ω^2 vanishes and, therefore, contains the stability boundary. Since the eigenvalues ω^2 are always real, it follows that *flutter instability cannot occur, and the only type of instability exhibited by a conservative system is divergence.* It is also noted that while the characteristic surfaces are defined in the $\omega^2 - \eta_k$ space, the stability and divergence boundaries lie in the η_k subspace.

It will now be shown that the fundamental characteristic surface is a *strictly convex surface* which cannot have convexity toward the fundamental region. This can be proved in different ways, and three methods will be discussed in this and the following sections since each will shed some light on different aspects of the behaviour of the system.

Noting again that both M and K are symmetric while M is also positive definite, a linear nonsingular transformation can be introduced to transform the second order equation (3.2) into a canonical form which results in n uncoupled equations. Thus, introduce

$$q = Tp, \qquad |T| \neq 0 \tag{3.12}$$

such that M and K are simultaneously diagonalized (Section 1.3), and (3.2) takes the form, on pre-multiplying by T',

$$\ddot{p} + W^2 p = 0 \quad \text{where} \quad W^2 = \text{diag}\,[\omega_1^2, \omega_2^2, \ldots \omega_n^2] \tag{3.13}$$

or

$$\ddot{p}_j + \omega_j^2 p_j = 0, \qquad j = 1, 2, \ldots, n,$$

with the associated characteristic equation

$$|W^2 - \omega^2 I| = 0 \tag{3.14}$$

The p_j are called *principal coordinates* of the system, and it is observed that the motion associated with one principal coordinate is independent of the remaining coordinates.

Arranging the eigenvalues ω_j^2 in the ascending order

$$\omega_1^2 \leq \omega_2^2 \leq \cdots \leq \omega_n^2, \tag{3.15}$$

it is readily shown on the basis of (3.14) that for $\eta_k = \eta_k^0$ and any ω_0^2 for which

$$\omega_0^2(\eta_k^0) < \omega_1^2(\eta_k^0), \tag{3.16}$$

the matrix

$$[W^2 - \omega_0^2 I],$$

and consequently the matrix

$$[-\omega_0^2 M + U - \eta_k^0 E^k], \tag{3.17}$$

is positive definite. With reference to Figure 3.1(b) and (c), $\omega_1^2 = \omega_1^2(\eta_k)$ represents the fundamental characteristic surface, and $\omega_1^2(\eta_k) = 0$ describes the stability boundary. For all points in the η_k space lying inside this boundary the system is stable and all ω_j^2's, including ω_1^2, are positive. Points outside the boundary yield an unstable system with one or more of the ω_j^2's negative. Although such points may not have practical significance it is possible to present a more complete theory by thinking in mathematical terms only.

3.2 *Convexity of the fundamental characteristic surface*

To demonstrate the convexity of the fundamental characteristic surface [29], consider an arbitrarily chosen point (ω_0^2, η_k^0) *in the fundamental region* and a ray τ (Fig. 3.1(b)) defined by

$$\begin{aligned} \eta_k &= \eta_k^0 + l_k\tau, \qquad k = 1, 2, \ldots, m, \\ \omega^2 &= \omega_0^2 + l\tau, \end{aligned} \tag{3.18}$$

emerging from this point in an arbitrarily chosen direction determined by the direction cosines (l_k, l). The radius vector τ is only allowed to take positive values. Introducing (3.18) into (3.11) yields

$$\Delta = |A(\eta_k^0, \omega_0^2) - \tau B(l_k, l)| \tag{3.19}$$

where

$$A = -M\omega_0^2 + U - \eta_k^0 E^k$$

and

$$B = l_k E^k + lM.$$

Since the matrix A is positive definite and both A and B are symmetric, it is again possible to introduce a congruence transformation (Section 1.3) to diagonalize A and B simultaneously. Thus, supposing that Q is the nonsingular matrix to effect such a reduction one has

$$Q'AQ - \tau Q'BQ = I - \tau C \tag{3.20}$$

where C is diagonal with the elements c_i. The determinant of (3.20) vanishes if (3.19) vanishes; hence, the variation of

$$\Delta = |I - \tau C|$$

or

$$\Delta = \begin{vmatrix} 1-\tau c_1 & 0 & \cdots \\ 0 & 1-\tau c_2 & \cdots \\ \cdots & \cdots & \cdots \end{vmatrix} \tag{3.21}$$

along the ray must be considered. First, note that the $c_i (i = 1, 2, \ldots, n)$ can take positive or negative values depending on the chosen direction of the ray (as determined by the direction cosines l_k and l). For $\tau = 0$ the system is stable as expected since the ray was directed from a point within the fundamental region. Proceeding now along the ray, the fundamental characteristic surface may or may not be intersected depending on the direction of the ray. In the latter case, all the c_i are negative and the determinant (3.21) remains positives as τ increases. If at least one c_i, say c_k, is positive, then proceeding along the ray by increasing the positive quantity τ from zero, the stage at which

$$1-\tau c_k=0$$

will eventually be reached, resulting in a zero determinant. When this happens, the ray intersects the fundamental characteristic surface. If more than one c_i is positive, intersections with other characteristic surfaces, associated with corresponding coordinates, are encountered as τ is increased. Assuming that c_k is the largest positive c_i, it is observed that $1-\tau c_k<0$ once the fundamental region is left. Clearly, a ray in an arbitrary direction emerging from an arbitrary point in the fundamental region cannot intersect a characteristic surface associated with a principal coordinate more than once, and if the fundamental region is once left by proceeding along the ray it cannot be reentered.

The obvious result of these arguments can be formulated [29] as a theorem:

THEOREM 3.1. *The fundamental characteristic surface cannot have convexity toward the fundamental region.*

An immediate corollary of this theorem is concerned with the convexity of the stability boundary which is a part of the fundamental characteristic surface. Due to its significance, however, it will be stated here as another theorem:

THEOREM 3.2. *The stability boundary cannot have convexity toward the region of stability.*

It seems that this theorem was first proved by Papkovich much earlier than the publication in Ref. [30] indicates. A similar result was obtained by Buckens [31] who studied linear eigenvalue problems. A generalization of Theorem 3.2 to systems exhibiting linear pre-critical deformations can be found in Refs. [32, 10]. Masur [33] showed that Theorem 3.1, and hence Theorem 3.2, remains valid for continuous systems as well.

It may be useful to know under what conditions the characteristic surface consists of planes, and it is readily deduced from (3.11) that if the matrices M, U and E^k can all be brought to a diagonal form simultaneously, then the surfaces described by (3.11) are indeed planes (hyperplanes). In particular, if the system has only one degree of freedom, obviously there can be no curved surfaces. Thus, the following theorem can be formulated:

THEOREM 3.3. *If the matrices M, U and E^k can be reduced to a diagonal* *form simultaneously, then the characteristic surface (and hence the stability boundary) consists of planes.*

3.3 Application of Rayleigh quotient

The extremum properties of the Rayleigh quotient provide a convenient tool for establishing bounds on eigenvalues, and will now be used to demonstrate the convexity of the fundamental characteristic surface through another approach.

Consider again the eigenvalue problem

$$(-\omega^2 M + U - \eta_k E^k)u = 0; \tag{3.22}$$

for fixed values of η_k, the frequencies of free vibration can be expressed as

$$\omega^2 = \frac{\langle u, (U - \eta_k E^k)u\rangle}{\langle u, Mu\rangle}, \tag{3.23}$$

Setting $\eta_k = 0$, one obtains the *natural frequencies*

$$\omega^2 = \frac{\langle u, Uu\rangle}{\langle u, Mu\rangle} \tag{3.24}$$

which are obviously positive.

It is known from Section 1.5 that the eigenvalues $\omega_i^2 (i = 1, 2, \ldots, n)$ are the stationary values of the quotient

$$R(x) = \frac{\langle x, (U - \eta_k E^k)x\rangle}{\langle x, Mx\rangle} \tag{3.25}$$

for any given set of η_k, where x is an arbitrary real vector. Also, if the eigenvalues are in the ascending order (3.15) one has the inequalities

$$\omega_1^2 \leq \frac{\langle x, (U - \eta_k E^k)x\rangle}{\langle x, Mx\rangle} \leq \omega_n^2. \tag{3.26}$$

Similarly, the divergence values of the parameters η_k can be expressed as the stationary values of an appropriate quotient. Consider, for example, the *ray*

$$\eta_k = l_k \xi, \tag{3.27}$$

emerging from the origin of the η_k space (Fig. 3.1(c)), where l_k and ξ are defined as in the case of τ except that ξ is now allowed to take negative as well as positive values, actually measuring the distance from the origin in both directions. The divergence values of ξ, characterized by $\omega^2=0$, are then obtained as

$$\langle u,(U-\xi l_k E^k)u\rangle=0 \tag{3.28}$$

or

$$\frac{1}{\xi_j}=\frac{\langle u^j, Eu^j\rangle}{\langle u^j, Uu^j\rangle} \quad \text{where} \quad j=1,2,\ldots,n \quad \text{and} \quad E=l_k E^k. \tag{3.29}$$

Assuming that the inequalities

$$\frac{1}{\xi_1}\geq\frac{1}{\xi_2}\geq\cdots\geq\frac{1}{\xi_n} \tag{3.30}$$

hold, the stationary values of the quotient

$$R(x)=\frac{\langle x, Ex\rangle}{\langle x, Ux\rangle} \tag{3.31}$$

yield $1/\xi_j$ $(j=1,2,\ldots,n)$ such that

$$\frac{1}{\xi_1}\geq\frac{\langle x, Ex\rangle}{\langle x, Ux\rangle}\geq\frac{1}{\xi_n}. \tag{3.32}$$

Consider now an arbitrary point $C(\omega_0^2,\ \eta_k^0)$ on the fundamental characteristic surface (Fig. 3.2) and denote the corresponding eigenvector by x. It can be shown that the equation

$$\langle x, (-\omega^2 M+U-\eta_k E^k)x\rangle=0 \tag{3.33}$$

describes a *plane* tangent to the fundamental surface at the point $(\omega_0^2,\ \eta_k^0)$. First, note that (3.33) does, indeed, describe a plane in $\omega^2-\eta_k$ space,

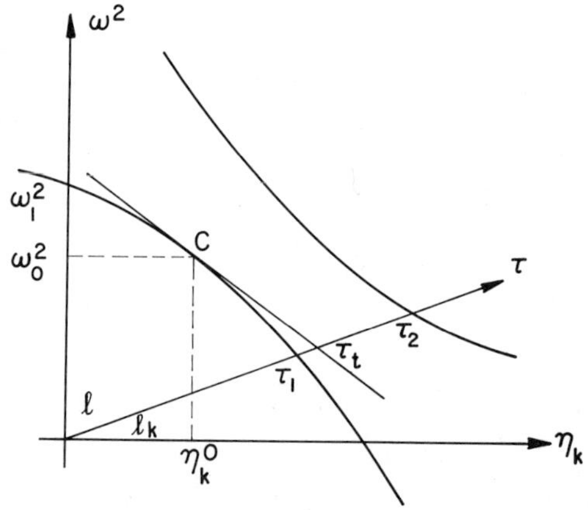

Figure 3.2

and that (ω_0^2, η_k^0) satisfies this equation by virtue of (3.22), x being the corresponding eigenvector. To show that (3.33) is tangent to the fundamental characteristic surface, suppose the latter is given in the form $\omega_1^2 = \omega_1^2(\eta_k)$ in the vicinity of the point C, where $k = 1, 2, \ldots, m$ This assumed solution should then satisfy (3.22) identically, which can be written as

$$[-\omega_1^2(\eta_k)M + U - \eta_k E^k]u^1 = 0. \tag{3.34}$$

Differentiating (3.34) with respect to any η_k yields

$$\left(-\frac{\partial \omega_1^2}{\partial \eta_k} M - E^k\right)u^1 + (-\omega_1^2 M + U - \eta_k E^k)\frac{\partial u^1}{\partial \eta_k} = 0$$

which upon pre-multiplying by u^1 and evaluating at C results in

$$\left\langle x, \left(-\frac{\partial \omega_1^2}{\partial \eta_k} M - E^k\right)x\right\rangle = 0, \tag{3.35}$$

giving the slopes of the fundamental characteristic surface as

$$\left.\frac{\partial \omega_1^2}{\partial \eta_k}\right|_C = -\frac{\langle x, E^k x\rangle}{\langle x, Mx\rangle}, \qquad k = 1, 2, \ldots, m. \tag{3.36}$$

It is immediately observed that the plane (3.33) has the same slopes and it is, therefore, tangential to the fundamental characteristic surface at C.

Consider next the ray

$$\left.\begin{aligned} \eta_k &= l_k \tau \\ \omega^2 &= l\tau \end{aligned}\right\} \tag{3.37}$$

passing through the origin of the $\omega^2 - \eta_k$ space and intersecting the tangent plane (3.33) as well as the characteristic surfaces. Here l_k and l are the direction cosines and τ measures the distance from the origin, assuming negative as well as positive values. The intersection of the ray (3.37) and the plane (3.33) yields (Fig. 3.2)

$$\tau_t = \frac{\langle x, Ux\rangle}{\langle x, (M_1 + E)x\rangle} \quad \text{or} \quad \frac{1}{\tau_t} = \frac{\langle x, (M_1 + E)x\rangle}{\langle x, Ux\rangle} \tag{3.38}$$

where $M_1 = lM$ and $E = l_k E^k$.

On the other hand, the intersections of the ray (3.37) with the characteristic surfaces result in

$$\tau_j = \frac{\langle u^j, Uu^j\rangle}{\langle u^j, (M_1 + E)u^j\rangle} \quad \text{or} \quad \frac{1}{\tau_j} = \frac{\langle u^j, (M_1 + E)u^j\rangle}{\langle u^j, Uu^j\rangle} \tag{3.39}$$

where $j = 1, 2, \ldots, n$. In analogy with (3.30), assuming that

$$\frac{1}{\tau_1} \geq \frac{1}{\tau_2} \geq \cdots \geq \frac{1}{\tau_n}$$

holds, the extremum properties of the Rayleigh quotient yield

$$\frac{1}{\tau_1} \geq \frac{1}{\tau_t} \quad \text{or} \quad \tau_1 \leq \tau_t \tag{3.40}$$

since $1/\tau_1$ is the largest eigenvalue associated with U and $M_1 + E$.

Recalling that the point C and the ray (3.37) were chosen arbitrarily, Theorem 3.1 follows.

Normality condition

Differentiating (3.22) along any smooth path s on the fundamental characteristic surface and pre-multiplying by $(u^1)'$ results in

$$\langle u^1, (\dot{\omega}_1^2 M + \dot{\eta}_k E^k) u^1 \rangle = 0 \tag{3.41}$$

where the dots denote differentiation with respect to s. (3.41) can be written as the inner product

$$\langle \dot{\rho}, \varepsilon \rangle = 0 \tag{3.42}$$

where the vectors $\dot{\rho}$ and ε are given as

$$\left.\begin{aligned} &\dot{\rho} = (\dot{\omega}_1^2, \dot{\eta}_1, \dot{\eta}_2, \ldots, \dot{\eta}_k) \\ &\text{and} \\ &\epsilon = (\langle u^1, Mu^1 \rangle, \langle u^1, E^1 u^1 \rangle, \ldots, \langle u^1, E^m u^1 \rangle)'. \end{aligned}\right\} \tag{3.43}$$

Equation (3.42) represents a *normality condition* in the sense that the vector ε is orthogonal to the plane tangent to the fundamental characteristic surface [34].

The intersections of the planes $\omega_1^2 = \text{const.}$ with the fundamental characteristic surface form certain *contours*, the stability boundary being associated with $\omega_1^2 = 0$. In this case the term with $\dot{\omega}_1^2$ in (3.41) vanishes and the vectors in (3.43) are given by

$$\left.\begin{aligned} &\dot{\rho} = (\dot{\eta}_1, \dot{\eta}_2, \ldots, \dot{\eta}_k) \\ &\text{and} \\ &\epsilon = (\langle u^1, E^1 u^1 \rangle, \langle u^1, E^2 u^1 \rangle, \ldots, \langle u^1, E^m u^1 \rangle)'. \end{aligned}\right\} \tag{3.44}$$

This normality condition which is related to the subspace η_k was first given by Masur [33] for continuous systems. It is analogous to the well-known normality condition associated with *yield surfaces* in Plasticity [35]. A similar normality condition was shown [36] to exist in buckling and post-buckling problems of certain types of structures in the space formed by all statically indeterminate self-equilibrated stress fields.

The normality condition (3.42) may also be used to prove Theorem 3.1. Thus, considering the ray

$$\rho = \rho_0 + b\dot{\rho}, \qquad b = \text{const.}, \tag{3.45}$$

in vector form, which lies on the plane tangent to the fundamental characteristic surface at a point $\rho_0(\omega_0^2, \eta_k^0)$, the quadratic form

$$f = \langle x, (-\omega^2 M + U - \eta_k E^k)x\rangle, \tag{3.46}$$

where x is the eigenvector associated with ρ_0, takes the form

$$f = \langle x, (-\omega_0^2 M + U - \eta_k^0 E^k)x\rangle + b\langle x, (-\dot{\omega}^2 M + U - \dot{\eta}_k E^k)x\rangle \tag{3.47}$$

along the ray (3.45). It follows that $f = 0$ since both the first and second terms in (3.47) vanish. The first term vanishes because (ω_0^2, η_k^0) is on the characteristic surface and the second term vanishes by virtue of the normality condition (3.42). Clearly, the matrix of (3.46) is not positive definite along the ray (3.45) which cannot, therefore, enter the fundamental region. This result could, of course, be readily deduced from the equation of the tangent plane given by (3.33) which indicates clearly that the quadratic form (3.46) is not positive definite along the tangent plane, and hence, the latter must lie totally outside the fundamental region.

3.4 Practical implications

The theorems established can be used either to estimate the frequencies of free vibration at certain values of the parameters η_k without necessarily repeating the analysis each time or, alternatively, to estimate the stability boundary of the system.

In the former case, suppose that both the stability boundary and the first natural frequency at $\eta_k = 0$ are known. On the basis of Theorem 3.1 it is then possible to establish a *lower bound* on the frequencies of free vibration, for any combination of the parameters η_k lying on the safe side of the stability boundary and in the positive η_k domain, as a surface generated by a straight line passing through the known point on the ω^2-axis and tracing the stability boundary. This is only one way of constructing a lower bound; other possibilities exist. For instance, one might establish the critical value for each $\eta_k (k = 1, 2, \ldots, m)$ individually in addition to the fundamental natural frequency at $\eta_k = 0$. A plane

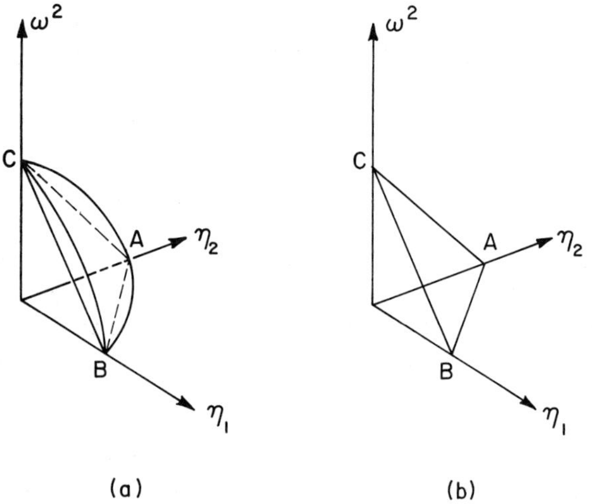

Figure 3.3

(hyperplane) may then be constructed as a lower bound to the fundamental characteristic surface. For a system with $m=2$, the lower bound by this procedure takes the form of a plane passing through points A, B and C (Fig. 3.3). Clearly, this last procedure yields a lower bound which is not as close as that obtained by the first procedure.

If the system has only one degree of freedom, or the equations uncouple as Theorem 3.3 hypothesizes, the lower bounds just indicated coincide with the fundamental characteristic surface as shown in (Fig. 3.3(b)).

If in addition to the first frequency the higher order frequencies are also required, they can be determined for each given set of η_k^0 by making use of the theory leading to the first proof of Theorem 3.1. Setting $l_k=0$, $\omega_0^2=0$, and $l=1$ in the ray (3.18), i.e., directing a ray from the point η_k^0 in the η_k space parallel to the ω^2-axis such that $\omega^2=\tau$, one observes from (3.18) and (3.19) that

$$|-\omega^2 M+U-\eta_k^0 E^k|=0$$

which, upon diagonalization, eventually reduces to (3.14), giving the n roots $\omega_i^2=\omega_i^2(\eta_k^0)$. If η_k^0 represents a point in the region of stability, all ω_i^2 are positive. It must be noted that although in the majority of systems with several degrees of freedom the frequencies are usually distinct, it may nevertheless happen that several frequencies coincide or become nearly coincident. In these cases, the theorems gain even more significance.

In the special case of proportional parametric dependence characterized by $\eta_k = l_k\xi$, one has only one independent parameter, ξ, and the characteristic surface

$$|-\omega^2 M + U - \xi E| = 0$$

reduces to a straight line, hyperbola, or other second order curve depending on whether the system has one, two, or more degrees of freedom.

The second proposed use of the theorems, namely, the estimation of the stability boundary of the system, will now be discussed. It is first recalled that the *divergence surfaces* (divergence boundary) are obtained from the characteristic surfaces by setting $\omega^2 = 0$, and the stability boundary is part of the fundamental characteristic surface.

Suppose the system has one independent parameter η_1 or more generally ξ as defined earlier; then, if the fundamental frequencies $\omega_{11}^2(\xi_1)$ and $\omega_{12}^2(\xi_2)$ are determined at two selected values of ξ, an upper bound to the critical value of ξ can immediately be obtained as the intersection with the ξ-axis of the straight line joining the points (ω_{11}^2, ξ_1) and (ω_{12}^2, ξ_2). This is illustrated in Figure 3.4. The upper-bound estimate may be considerably improved by careful selection of ξ_1 and ξ_2. It appears that the closer these are to ξ_{cr}, the better the estimate.

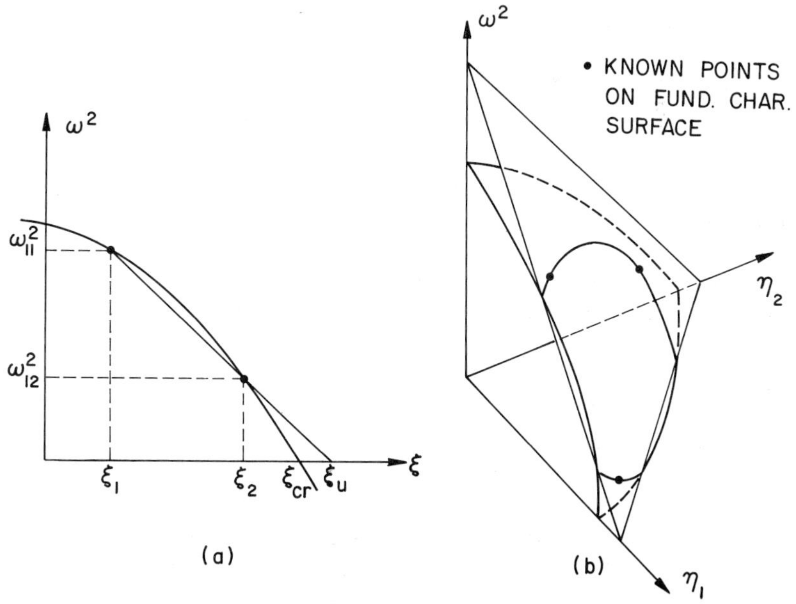

Figure 3.4

The case of a system with more than one independent parameter η_k presents some complications, however. It is, of course, theoretically possible to construct a tangent (hyper) plane to the fundamental characteristic surface at any *regular* point on this surface and regard its intersection with the η_k subspace as an upper bound to the stability boundary. This upper bound, however, will not be a close one in general. Alternatively, if the system has m independent parameters η_k, by establishing $m+1$ points on the fundamental characteristic surface a plane may be constructed whose intersection with the η_k subspace provides another estimate to the stability boundary. There is no gaurantee, however, that this will be an upper bound to the whole stability boundary. This fact is illustrated in Figure 3.4(b) for $m=2$.

A close upper bound on the whole stability boundary may be obtained by repeating the operation illustrated in Figure 3.4(a) a sufficient number of times for suitably different rays $\eta_k = l_k\xi$ in the η_k space. The individual upper bounds ξ_u thus obtained may then be joined to form a piecewise linear upper bound made up of straight lines for $m=2$, planes for $m=3$ and hyperplanes for $m>3$.

Clearly, any one of the methods described here will yield the exact stability boundary when the system obeys Theorem 3.3.

The practical usefulness of these observations lies in the ability to estimate the stability boundary experimentally through non-destructive vibration tests. For example, a thin plate of completely arbitrary shape and edge conditions, with independent in-plane loads can be tested for the fundamental frequency at a number of load combinations. An estimate of the stability boundary can thus be obtained without resorting to a difficult analytical stability analysis. It is understood that a continuous plate can effectively be reduced to a system with finite degrees of freedom through finite element or Rayleigh-Ritz procedures, thus reducing the potential energy to a quadratic form and bringing the problem into the scope of the present formulation.

3.5 Effect of damping

If damping is added to the conservative system considered in the preceding section, the system ceases to be conservative. Consider, for example, damping forces proportional to generalized velocities and derivable from the Rayleigh dissipation function

$$R = \tfrac{1}{2}\langle \dot{q}, D\dot{q}\rangle; \tag{3.48}$$

the differential equations of motion then take the form

$$M\ddot{q}+D\dot{q}+Kq=0 \tag{3.49}$$

upon applying Lagrange's equations to (3.1) and (3.48). Here D is a non-negative definite matrix. Alternatively, (3.49) can be expressed as a first order equation of the form

$$\dot{x}=Ax \tag{3.50}$$

where

$$A=\begin{bmatrix}-M^{-1}D & -M^{-1}K\\ \mathrm{I} & 0\end{bmatrix}, \qquad x=\begin{bmatrix}\dot{q}\\ q\end{bmatrix}.$$

Assuming solutions of the form

$$q=ue^{\lambda t},$$

(3.49) yields

$$(M\lambda^2+D\lambda+K)u=0 \tag{3.51}$$

or equivalently

$$(A-\lambda I)z=0 \quad \text{where} \quad z=\begin{bmatrix}\lambda u\\ u\end{bmatrix}. \tag{3.52}$$

The stability of damped systems described by (3.49) is discussed by several authors [25, 37–41]. The first question that comes to mind is, of course, concerned with the stability of the damped system when the corresponding conservative system is stable, and it can readily be shown that the former will be asymptotically stable if the latter is stable and the matrix D is positive definite. Thus, note that the Hamiltonian (3.3) is again a Lyapunov function, because both T and K are positive definite, and

$$\dot{H}=-\langle\dot{q},D\dot{q}\rangle<0, \tag{3.53}$$

which follows from

$$\dot{H}=\langle\dot{q},M\ddot{q}\rangle+\langle\dot{q},Kq\rangle$$

and (3.49), indicating stability.

Next, pre-multiply (3.51) by $\bar{u}$ to obtain

$$\lambda^2+d\lambda+k=0 \tag{3.54}$$

where u is chosen such that $\langle \bar{u}, Mu \rangle = 1$, while $d = \langle \bar{u}, Du \rangle > 0$, $k = \langle \bar{u}, Ku \rangle > 0$ (see Section 1.2). (3.54) yields

$$\lambda = \frac{-d \pm (d^2 - 4k)^{\frac{1}{2}}}{2} \tag{3.55}$$

where the absolute value of d is always greater than that of $(d^2 - 4k)^{\frac{1}{2}}$ when the latter is real. It follows that λ will always have a negative real part and hence the equilibrium is asymptotically stable provided $k > 0$.

If D is positive semi-definite, the equilibrium state $q = \dot{q} = 0$ can still be asymptotically stable. In other words, the positive definiteness of D is sufficient but not necessary for asymptotic stability, and the roots of (3.51) can all have negative real parts even when D is non-negative definite but not positive definite. The necessary and sufficient conditions for the asymptotic stability when $D \geq 0$ can be derived in different ways. If a particular system is given, the most direct way of ascertaining whether the system is asymptotically stable is to determine the $2n$ roots of (3.51) and see if all have negative real parts. In some cases the Routh-Hurwitz criterion or the Lyapunov equation based on (3.50) may be easier to apply (see Section 2.3). Recently, Moran [38] and Walker and Schmitendorf [39] gave other methods.

Suppose, for some eigenvalue λ, $d = 0$ and λ is, therefore, imaginary as can be seen from (3.55). Since D is positive semi-definite, $d = 0$ implies

$$Du = 0. \tag{3.56}$$

Introducing (3.56) into (3.51), it is observed that u is also an eigenvector of the conservative system

$$(M\lambda^2 + K)u = 0. \tag{3.57}$$

Conversely, if u is an eigenvector of the above conservative system and it satisfies $Du = 0$, then u is also an eigenvector of (3.51) and the corresponding eigenvalue λ is imaginary. Thus, the necessary and sufficient conditions that the system (3.51) be asymptotically stable are that *none of the eigenvectors of the conservative system* (3.57) *lies in the null space of D* (see Ref. [38]). Systems satisfying these conditions are called *completely damped* (*pervasively* or *strictly damped*).

In order to establish certain bounds for the roots of the damped system (3.51), write

$$\lambda = \gamma + i\omega, \qquad i = \sqrt{-1}. \tag{3.58}$$

where γ and ω are real, and introduce (3.58) into (3.54) to obtain

$$\gamma^2 - \omega^2 + d\gamma + k = 0$$

and

$$2\gamma\omega + d\omega = 0, \tag{3.59}$$

upon equating real and imaginary parts of the left-hand side of (3.54) to zero. The second equation in (3.59) yields

$$\omega = 0 \quad \text{or} \quad \gamma = -\frac{d}{2}. \tag{3.60}$$

Using $\omega = 0$ in the first equation gives

$$\gamma = \frac{-d \pm (d^2 - 4k)^{\frac{1}{2}}}{2} \tag{3.61}$$

which indicates that $d^2 - 4k \geq 0$ since γ is real. It follows that the real roots are all negative provided $k > 0$. When $k = 0$ the undamped system exhibits a divergence point, and it follows from (3.61) that so does the damped system, since γ as well as ω vanishes. In fact, it is clear from (3.51) that whenever $\lambda = 0$, indicating divergence for the damped system, the characteristic equation takes the form

$$|K| = 0$$

which shows that the undamped system also exhibits a divergence point simultaneously. It follows that the stability boundary of the conservative system remains unaffected in the presence of damping, and the theorems of Sections 3.4 are valid for the damped system as well.

If the conservative system is unstable, then the matrix K is no longer a definite matrix, and k can take on negative values resulting in a positive γ and hence instability. Conversely, if $\gamma > 0$ it follows from (3.61) that $k < 0$. It is important to note that the root λ can only have positive real part if $\omega = 0$, that is, if it is actually real. This can be deduced from (3.55) and (3.61), and verified by considering the case when $\omega \neq 0$. Thus, from (3.60), one has $\gamma = -d/2 < 0$, and after substitution into the first equation of (3.59),

$$k = \omega^2 + \gamma^2 = |\lambda|^2 \tag{3.62}$$

which indicates that the conservative system is also stable. As a result of these arguments one concludes that if nonnegative definite damping is added to a conservative system, the resulting dissipative system *cannot exhibit flutter instability*. Such systems, like the undamped conservative systems, can only lose stability by divergence.

It also follows from Section 1.5 (see Exercises 2 and 3) that

$$\omega_{10}^2 \leq k \leq \omega_{n0}^2 \tag{3.63}$$

where w_{i0}^2 $(i=1, 2, \ldots, n)$ are the eigenvalues of the undamped conservative system. Similarly,

$$\mu_1 \leq d \leq \mu_n \tag{3.64}$$

where

$$\mu_1 \leq \mu_2 \leq \cdots \leq \mu_n$$

and μ_i are the eigenvalues of the matrix pencil

$$-\mu M + D.$$

It is then inferred from (3.62) and (3.63) that for $\omega \neq 0$

$$\omega_{10} \leq |\lambda| \leq \omega_{n0} \tag{3.65}$$

and from (3.60) and (3.64)

$$-\tfrac{1}{2}\mu_n \leq \gamma \leq -\tfrac{1}{2}\mu_1. \tag{3.66}$$

Inequalities (3.65) and (3.66) prescribe a bounded domain in the complex λ-plane which contains all the complex roots. For further discussion concerning the bounds on eigenvalues of the damped system see Refs [37] and [40].

3.6 Example: A two-degree-of-freedom model

To illustrate some of the concepts introduced in this chapter, consider the rigid link structural model, Figure 3.5, consisting of two pin-jointed, straight

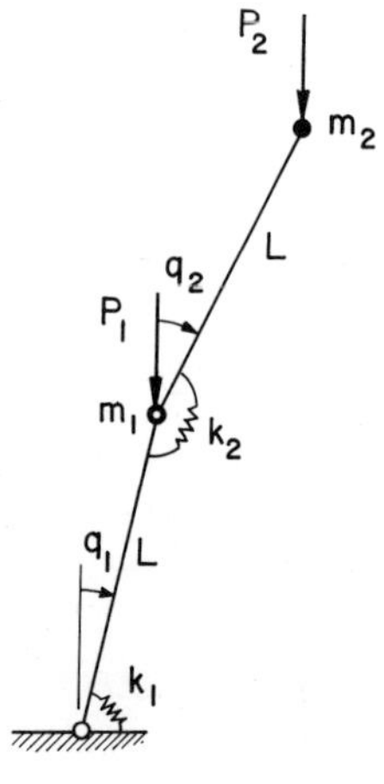

Figure 3.5

members. Each of the two joints has a linear torsional spring of stiffness $k_1 = 2k$ and $k_2 = k$. Two independent concentrated loads p_1 and p_2 act as shown. Point masses $m_1 = 2m$ and $m_2 = m$ are attached to the otherwise weightless members as shown.

For a linear analysis the potential and kinetic energy can be written as

$$V = kq_1^2 + \tfrac{1}{2}k(q_2 - q_1)^2 - \tfrac{1}{2}Lp_1q_1^2 - \tfrac{1}{2}Lp_2(q_1^2 + q_2^2),$$
$$T = \tfrac{1}{2}mL^2(3\dot{q}_1^2 + \dot{q}_2^2 + 2\dot{q}_1\dot{q}_2).$$

Introducing the nondimensional mass and load parameters

$$\bar{m} = \frac{mL^2}{k} \quad \text{and} \quad \eta_i = \frac{p_iL}{k},$$

and making use of Lagrange's equations yield the following characteristic equation:

$$\begin{vmatrix} -3\bar{m}\omega^2 + 3 - \eta_1 - \eta_2 & -\bar{m}\omega^2 - 1 \\ -\bar{m}\omega^2 - 1 & -\bar{m}\omega^2 + 1 - \eta_2 \end{vmatrix} = 0. \tag{3.67}$$

This equation defines the characteristic surfaces. Setting $\omega^2 = 0$ in (3.67), one obtains the *divergence surfaces* including the *stability boundary* in the η_k space.

The expanded version of (3.67) for $\bar{m} = 1$ is

$$2\omega^4 + \eta_2^2 + \omega^2\eta_1 + 4\omega^2\eta_2 + \eta_1\eta_2 - 8\omega^2 - \eta_1 - 4\eta_2 + 2 = 0 \tag{3.68}$$

and the corresponding equation for divergence surfaces is

$$\eta_2^2 + \eta_1\eta_2 - \eta_1 - 4\eta_2 + 2 = 0. \tag{3.69}$$

(3.69) represents a hyperbola in the η_1–η_2 plane as shown in Figure 3.6(a). Obviously, the stability boundary is concave toward the origin.

(3.68) is a second-degree equation (in ω^2), representing a quadric surface in the $\omega^2 - \eta_1 - \eta_2$ space. In fact, the surface consists of two parts, the one closest to the origin being the fundamental characteristic surface. By studying the intersections of (3.68) with the three coordinate planes this becomes evident. The intersection with the plane $\eta_2 = 0$ yields the hyperbola in Fig. 3.6 (b). The intersection with $\eta_1 = 0$ results in another hyperbola shown in Fig. 3.6 (c). Similarly, taking a ray $\eta_k = l_k\xi$ $(k = 1, 2)$ in the η_k space and substituting into (3.68) again results in a conic section in the $\omega^2 - \xi$ plane. This is equivalent to finding intersections of the surface with a plane containing, and rotating about, the ω^2-axis, its position being determined by the l_k $(k = 1, 2)$. The results will always

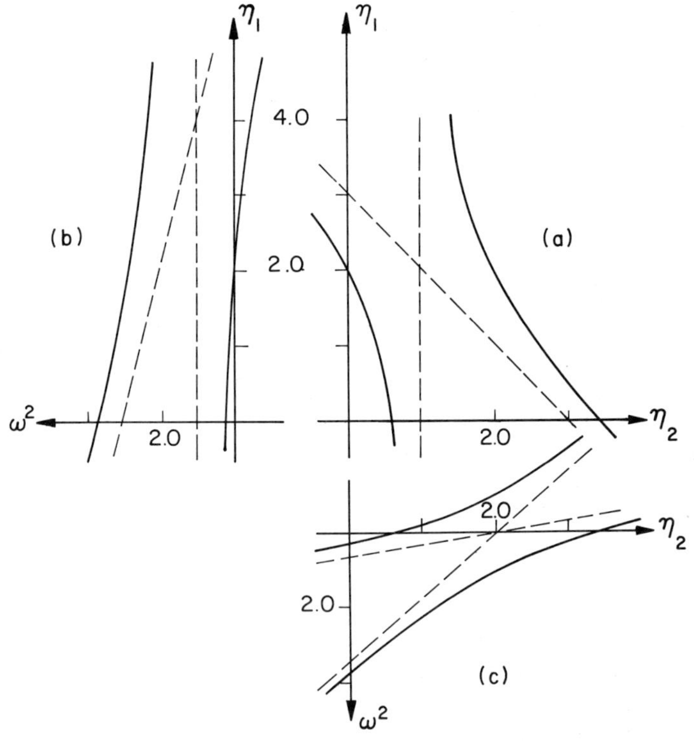

Figure 3.6

demonstrate that the fundamental characteristic curve is concave (if not a straight linc) toward the origin as the theorems predict.

It must be remarked here that although the example is chosen from structural mechanics, the theory of this chapter is applicable to problems in other disciplines as well.

3.7 Continuous systems

The rigorous analysis of a continuous system, which does not have finite degrees of freedom, may pose fundamental difficulties and may even be impossible. Methods of calculus of variations, for example, may not be practical besides leading only to the governing differential equation rather than the solution of the problem. There are, however, well-established methods which effectively discretize a given continuous body, thus resulting in a finite-degree-of-freedom system. Such a discretized model falls within the scope of the theory presented in this book.

Two approaches will be mentioned here, namely, the Rayleigh-Ritz method and Galerkin's method. The former approach is concerned with the energy of the system while the latter is often applied to the differential equation directly. Basically, one assumes a suitable shape for the deformation of the system, consisting of linear combinations of admissible functions. In the Ritz method these functions satisfy geometric boundary conditions only, while in Galerkin's method they are required to satisfy both the gometric and natural boundary conditions. These methods and the associated convergence properties are dealt with in several standard texts and will not be reiterated here.

The Rayleigh-Ritz procedure will now be illustrated in two examples.

1. *Euler Column*

Consider the axially-loaded and simply-supported elastic bar shown in Figure 3.7. The strain energy stored in the member due to bending is given by

$$U = \frac{1}{2}\int_0^l EI(y'')^2\,\mathrm{d}x \tag{3.70}$$

where E is the elastic modulus of the material and I is the moment of inertia of the cross section (constant along the bar). The work done by the applied axial force during bending is expressed by

$$\tfrac{1}{2}\eta\int_0^l (y')^2\,\mathrm{d}x. \tag{3.71}$$

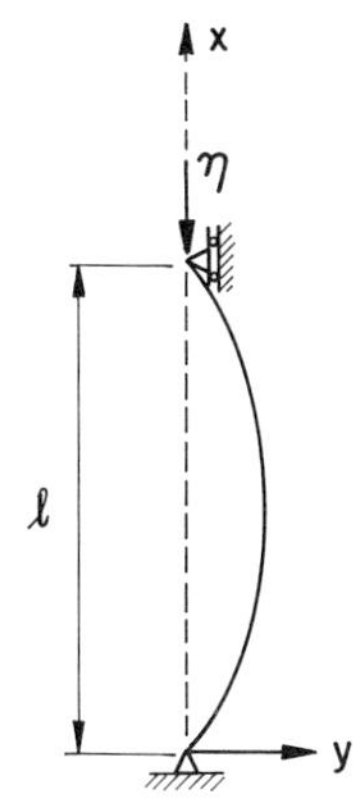

Figure 3.7

The total potential energy is then in the form

$$V=\tfrac{1}{2}EI\int_0^l (y'')^2\,dx-\tfrac{1}{2}\eta\int_0^l (y')^2\,dx \tag{3.72}$$

for small deformations.

Assuming now a deflection function in the form of the series

$$\begin{aligned} y&=q_1\sin\frac{\pi x}{l}+q_2\sin\frac{2\pi x}{l}+\cdots+q_n\sin\frac{n\pi x}{l}\\ &\equiv\sum_{k=1}^{n} q_k\sin\frac{k\pi x}{l}, \end{aligned} \tag{3.73}$$

one obtains

$$V=\frac{1}{2}\sum_{k=1}^{n}\frac{k^4\pi^4}{2l^3}EI\,q_k^2-\tfrac{1}{2}\eta\sum_{k=1}^{n}\frac{k^2\pi^2}{2l}q_k^2$$

where q_k are the generalized coordinates. The stationarity property yields

$$\frac{\partial V}{\partial q_k}=\left(\frac{k^4\pi^4}{2l^3}EI-\eta\frac{k^2\pi^2}{2l}\right)q_k=0;\qquad k=1,2,\ldots,n$$

which results in the Euler load for $k=1$,

$$\eta_{cr}=\frac{\pi^2 EI}{l^2}. \tag{3.74}$$

Note that in this particular example, the chosen deflection function is such that the continuous system is completely uncoupled and the stability determinant is in a diagonal form. Clearly, if the series (3.73) is allowed to be infinite the Euler load (3.74) will not change and it is, therefore, exact. The buckling mode corresponding to (3.74) is given by

$$y=q_1\sin\frac{\pi x}{l}. \tag{3.75}$$

2. *A column under its own weight*

Suppose now that the same bar is subjected to uniformly distributed constant-directional gravitational forces as shown in Figure 3.8. The total potential energy of this system is given by

$$V=\tfrac{1}{2}EI\int_0^l (y'')dx-\tfrac{1}{2}\eta\int_0^l (l-x)(y')^2\,dx \tag{3.76}$$

where η is the intensity of the applied force. The second integral concerning the work done by the applied force follows from the integral

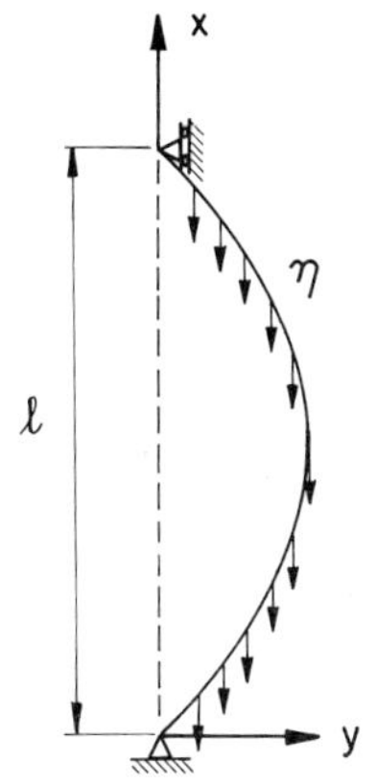

Figure 3.8

of the work of the elemental force $\eta\,\mathrm{d}x$ at a distance x which can be written as

$$\int_0^l \tfrac{1}{2}\eta\,\mathrm{d}x \int_0^x (y')^2\,\mathrm{d}x.$$

As a first approximation, assume a deflection function of the form

$$y = q\sin\frac{\pi x}{l}$$

which is the first term of the sine series (3.73). Introducing this function into the potential energy expression (3.76), one has

$$V = \tfrac{1}{2}EI\int_0^l q^2\frac{\pi^4}{l^4}\sin^2\frac{\pi x}{l}\,\mathrm{d}x - \tfrac{1}{2}\eta\int_0^l q^2(l-x)\frac{\pi^2}{l^2}\cos^2\frac{\pi x}{l}\,\mathrm{d}x$$

$$= \tfrac{1}{2}\left(\frac{\pi^4}{2l^3}EI\right)q^2 - \tfrac{1}{2}\eta\frac{\pi^2}{4}q^2 \tag{3.77}$$

and hence

$$\frac{\partial V}{\partial q} = \left(\frac{\pi^4}{2l^3}EI - \eta\frac{\pi^2}{4}\right)q = 0, \tag{3.78}$$

yielding the approximate critical load

$$\eta_{\mathrm{cr}} = 2\frac{\pi^2 EI}{l^3} = 19.74\frac{EI}{l^3}. \tag{3.79}$$

Recalling the extremum properties of Rayleigh's quotient, it is noted that (3.79) is an upper bound to the actual critical load. In fact, the exact value of η_{cr} can be determined [42] as $\eta_{\mathrm{cr}} = 18.57\,EI/l^3$.

For a better approximation, let the defleection function be

$$y = q_1 \sin \frac{\pi x}{l} + q_2 \sin \frac{2\pi x}{l}. \tag{3.80}$$

Introducing this function into (3.76), one obtains

$$V = \frac{1}{2}\left(\frac{\pi^4}{2l^3} EI\right) q_1^2 + \frac{1}{2}\left(\frac{8\pi^4}{l^3} EI\right) q_2^2 - \tfrac{1}{2}\eta \left[\frac{\pi^2}{4} q_1^2 + 2\frac{20}{9} q_1 q_2 + \pi^2 q_2^2\right] \tag{3.81}$$

and hence

$$\begin{bmatrix} \frac{\pi^4}{2l^3} EI - \eta \frac{\pi^2}{4} & -\frac{20}{9}\eta \\ -\frac{20}{9}\eta & \frac{8\pi^4}{l^3} EI - \eta\pi^2 \end{bmatrix} \begin{bmatrix} q_1 \\ q_2 \end{bmatrix} = 0. \tag{3.82}$$

The eigenvalues can readily be obtained as

$$\eta_{1\mathrm{cr}} = 18.58 \frac{EI}{l^3}, \qquad \eta_{2\mathrm{cr}} = 105.22 \frac{EI}{l^3}. \tag{3.83}$$

It is observed that the upper bound $\eta_{1\mathrm{cr}}$ to the exact critical load is very close, and a two-degree-of-freedom model can, therefore, be considered quite satisfactory.

A continuous system can, of course, be studied from a dynamics point of view as well by reducing it to a discrete system. Frequencies of free vibration and characteristic surfaces can thus be obtained through a similar procedure. This, however, will be done for more general systems in the following chapters and the student is urged to formulate such a dynamical system at this stage in Problem 2.

Problems

1. Combining Examples 1 and 2, determine the stability boundary of an elastic column subjected to the combined action of its weight and an axial load. Show that the stability boundary cannot have convexity toward the region of stability. Establish bounds for the boundary.

2. Formulate the system of Problem 1 from a dynamics point of view. Derive the governing differential equation by taking into account inertia forces. Reduce the system to a two-degree-of-freedom model and determine the characteristic surface (see Section 4.7).

4

Pseudo-conservative systems

This chapter is concerned with the stability properties of pseudo-conservative systems which contain nonconservative *circulatory forces* and are, therefore, essentially nonconservative. Nevertheless, as emphasized in Section 2.4, the physical properties of the systems in this class are such that they behave very much like conservative systems: the frequencies ω^2 of free vibrations are always real and a loss of stability occurs by divergence only, flutter instability being ruled out. Pflüger's rod [42], which consists of a simply-supported elastic bar subjected to uniformly-distributed tangential forces along its length, constitutes a typical example for this class. Leipholz [15] discussed the basic properties of these systems and demonstrated [43] the sufficiency of statical methods with particular emphasis on the energy criterion. A general theory concerning the free vibrations and stability of discrete multiple-parameter pseudo-conservative systems has been developed in Ref. [44], and one-dimensional continuous systems have been treated in Ref. [45]. Herrmann [46] and Sundararajan [47] have derived sufficient conditions for a system to be divergence type. Kornecki [48] discussed the character of instability of certain aero-elastic systems. It has further been shown [49] that, after an appropriate transformation, these systems lend themselves to familiar variational principles of conservative fields, and the extremum properties of the classical Rayleigh quotient are applicable.

4.1 Introduction

It is clear from the preceding chapter that the characteristic surface, i.e., the parameter-frequency relationship, is basic to an understanding of the instability behaviour as well as of the dynamic response of a system at given values of the parameters. General properties of the characteristic surface may also have significant practical implications, leading to lower and/or upper bounds for the stability boundary and/or frequencies, as in the case of conservative systems discussed in the preceding chapter.

The characteristic surface gains even more significance when nonconservative forces are present. The differing nature of instability behaviour of conservative and various classes of nonconservative systems can thus be illustrated conveniently. A nondissipative nonconservative system, for example, may lose stability by flutter when a branch of the characteristic curve (lying on the characteristic surface) reaches a maximum at which two initially distinct roots of the characteristic equation coalesce and, with a further increase of the associated parameter, become a pair of complex conjugate roots (Section 2.5), resulting in unbounded oscillations. Obviously, such a behaviour cannot be exhibited by a conservative system whose frequencies ω^2 are necessarily real and the characteristic surface does not have maxima with respect to parameters. In general, a nonconservative system can exhibit flutter as well as divergence instability, although the latter may sometimes occur before the former, a characteristic curve intersecting the parameter space before exhibiting a maximum. In this chapter, however, attention will be focussed on a class of nonconservative systems which is incapable of flutter at all. Mathematically, the degree of asymmetry of the matrices E^k associated with *circulatory* parameters plays a decisive role on the nature of instability behaviour, such that when the asymmetry is not strong enough the matrix $M^{-1}K$ is *symmetrizable* and the eigenvalues ω^2 remain real for all values of the parameters. In other words, the originally asymmetric matrix pencil can be transformed into a symmetric *simple* pencil with all its implications as regards to physical behaviour. Such systems can aptly be called *pseudo-conservative* and will now be studied in general terms.

4.2 Properties of the system

Consider again a linear, autonomous dynamical system described by the generalized coordinates $q_i (i = 1, 2, \ldots, n)$. The parameters $\eta_k (k = 1, 2, \ldots, m)$ can now represent *circulatory* as well as conservative forces. If the generalized forces F_i associated with the η_k are considered separately from the internal forces, then, to a first approximation,

$$F_i = \eta_k E^k q_i, \qquad k = 1, 2, \ldots, m \tag{4.1}$$

where the E^k corresponding to circulatory forces are asymmetric. Distinguishing further between the conservative and circulatory forces, one has

$$V = \tfrac{1}{2}\langle q, (U - \eta_\alpha E^\alpha) q\rangle \tag{4.2}$$

for conservative forces η_α where $\alpha = 1, 2, \ldots, m_1, m_1 < m$, and the remaining $m - m_l$ parameters represent the circulatory forces.

Assuming a quadratic form for the kinetic energy of the system as before, Lagrange's equations yield

$$M\ddot{q}+(U-\eta_k E^k)q=0 \tag{4.3}$$

or

$$M\ddot{q}+Kq=0 \tag{4.4}$$

where some E^k, and hence K, are asymmetric while $M=M'>0$.

The systems described by (4.4) will be called pseudo-conservative if the matrix

$$M^{-1}K \tag{4.5}$$

is *symmetrizable* (Section 1.4).

Assuming solutions of the form

$$q=ue^{i\omega t} \tag{4.6}$$

results in the eigenvalue problem

$$(-\omega^2 M+K)u=0 \quad \text{or equivalently} \quad (-\omega^2 I+M^{-1}K)u=0 \tag{4.7}$$

as in the case of conservative systems. Due to the asymmetry of K, however, the matrix pencil $-\omega^2 M+K$ is now associated with left eigenvectors as well as right eigenvectors which will be expressed as

$$v'(-\omega^2 I+M^{-1}K)=0$$

or (4.8)

$$(-\omega^2 I+K'M^{-1})v=0.$$

It was shown in Section 1.4 that a symmetrizable matrix is associated with a full complement of linearly independent eigenvectors and that the left and right eigenvectors are related by

$$v=T^2u \tag{4.9}$$

where T is positive definite and

$$T^2M^{-1}K=S_2 \quad \text{or} \quad TM^{-1}KT^{-1}=T^{-1}S_2T^{-1} \tag{4.10}$$

where S_2 is a symmetric matrix.

The characteristic equation

$$|-\omega^2 M+K|=0$$

yields n real eigenvalues ω^2 since $M^{-1}K$ is assumed to be symmetrizable, and (4.7) and (4.8) result in a full set of real right and left eigenvectors.

Using the *generalized Rayleigh quotient* (Section 1.5), one can express the eigenvalues

$$\omega^2 = \frac{\langle v, Ku \rangle}{\langle v, Mu \rangle} \tag{4.11}$$

as the stationary values of

$$R(x, y) = \frac{\langle y, Kx \rangle}{\langle y, Mx \rangle} \tag{4.12}$$

which does not, in general, have max-min properties. Since $M^{-1}K$ is symmetrizable, however, the eigenvalues ω^2 can also be expressed as the stationary values of the classical Rayleigh quotient. Thus, pre-multiplying (4.7) first by T^2 and then by u' results in

$$\omega^2 = \frac{\langle u, S_2 u \rangle}{\langle u, T^2 u \rangle} \tag{4.13}$$

upon using (4.10). Alternatively, post-multiplying the first equation in (4.8) by u and using (4.9) and (4.10) again leads to (4.13). The corresponding Rayleigh quotient

$$R(x) = \frac{\langle x, S_2 x \rangle}{\langle x, T^2 x \rangle} \tag{4.14}$$

satisfies the inequalities

$$\omega_1^2 \leq \frac{\langle x, S_2 x \rangle}{\langle x, T^2 x \rangle} \leq \omega_n^2, \tag{4.15}$$

assuming that the eigenvalues ω_i^2 are in the ascending order

$$\omega_1^2 \leq \omega_2^2 \leq \cdots \leq \omega_n^2.$$

Clearly, a system in the class under consideration can only lose stability by divergence, and since the pencil $-\omega^2 M + K$ is simple with linearly independent eigenvectors, the arguments concerning the stability of critical and noncritical equilibrium states of conservative systems (Section 3.1) are valid here as well and will not, therefore, be repeated. Different formulations of the system (4.4), however, are worth noting for future reference. Pre-multiplying (4.4), for example, first by M^{-1} and then by T^2, one obtains

$$T^2 \ddot{q} + S_2 q = 0 \tag{4.16a}$$

upon using (4.10). This equation is similar to that of a conservative system which has a fictitious mass matrix T^2 and potential energy matrix

S_2. Alternatively, introducing $q = T^{-1}r$ into (4.16a), pre-multiplying by T^{-1} and using (4.10), one obtains

$$\ddot{r} + Cr = 0 \tag{4.16b}$$

where $C = T^{-1}S_2T^{-1}$ is symmetric. Associated eigenvalue problems may be expressed as

$$(-\omega^2 I + C)y = 0, \tag{4.16c}$$

by introducing $r = ye^{i\omega t}$. The character of the eigenvalues ω^2 determines the stability properties of the system just like in conservative systems.

4.3 Convexity of the fundamental characteristic surface

It can be shown in analogy with conservative systems that the fundamental characteristic surface of pseudo-conservative systems cannot have convexity toward the fundamental region. This can be done in several ways following the methods of Sections 3.2 and 3.3; here, however, the proof will be carried out by means of the Rayleigh quotient.

Consider (4.7) in the explicit form [49]

$$[-\omega^2 I + M^{-1}(U - \eta_k E^k)]u = 0 \tag{4.17}$$

and the associated equation

$$v'[-\omega^2 I + M^{-1}(U - \eta_k E^k)] = 0. \tag{4.18}$$

Let $P(\omega_0^2, \eta_k^0)$ be an arbitrary regular point on the fundamental surface (Fig. 4.1), and denote the corresponding right and left eigenvectors by x and y respectively. It will be shown that the equation

$$\langle y, [-\omega^2 I + M^{-1}(U - \eta_k E^k)]x \rangle = 0 \tag{4.19}$$

describes a plane tangent to the fundamental surface at P.

It is clear that (4.19) does, indeed, represent a plane in the $\omega^2 - \eta_k$ space and that (ω_0^2, η_k^0) satisfies this equation. The slopes of this plane can be expressed as

$$\left.\frac{\partial \omega^2}{\partial \eta_k}\right|_P = -\frac{\langle y, M^{-1}E^k x\rangle}{\langle y, x\rangle}, \qquad k = 1, 2, \ldots, m. \tag{4.20}$$

To show tangentiality, differentiate (4.17) with respect to any η_k (see also Section 3.3) to get

$$\left(-\frac{\partial \omega^2}{\partial \eta_k} I - M^{-1}E^k\right)u + [-\omega^2 I + M^{-1}(U - \eta_k E^k)]\frac{\partial u}{\partial \eta_k} = 0$$

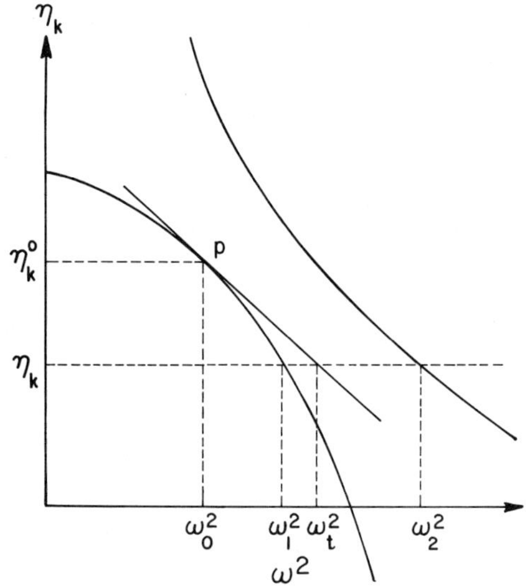

Figure 4.1

which upon pre-multiplying by v' and evaluating at P yields the derivatives (4.20). This completes the proof.

Instead of taking a ray from the origin, one can compare the eigenvalues ω^2 corresponding to arbitrary points in the η_k space. First, note that the equation of the tangent plane can be expressed as

$$\langle x, T^2[-\omega^2 I + M^{-1}(U - \eta_k E^k)]x\rangle = 0 \tag{4.21}$$

where T^2 is associated with the point P and the relation $y = T^2 x$ holds due to (4.9). Next, suppose η_k is an arbitrary given point in the η_k space; pre-multiplying (4.17) by T^2 and u', then, yields the eigenvalues in the form

$$\omega_i^2 = \frac{\langle u^i, T^2 M^{-1}(U - \eta_k E^k)u^i\rangle}{\langle u^i\, T^2 u^i\rangle}, \qquad i = 1, 2, \ldots, n. \tag{4.22}$$

It will be assumed that

$$\omega_1^2 \leq \omega_2^2 \leq \cdots \leq \omega_n^2 \tag{4.23}$$

holds. On the other hand, the intersection of the tangent plane (4.21) with the straight line parallel to the ω^2 axis and passing through the same arbitrary point η_k (Figure 4.1) results in the value $\omega^2 = \omega_t^2$ given by

$$\omega_t^2 = \frac{\langle x, T^2 M^{-1}(U - \eta_k E^k)x\rangle}{\langle x, T^2 x\rangle}. \tag{4.24}$$

Comparing (4.22) and (4.24) and using (4.15), one obtains

$$\omega_t^2 \geq \omega_1^2$$

which shows that the tangent plane has no common points with the *fundamental region.* Thus, one formulates the following theorems:

THEOREM 4.1. *The fundamental characteristic surface cannot have convexity toward the fundamental region.*

THEOREM 4.2. *The stability boundary cannot have convexity toward the region of stability.*

4.4 Lower bound estimates

It will be interesting to compare the behaviour of a pseudo-conservative system with that of *the corresponding conservative system* [44, 49]. The latter is defined by

$$(-\omega^2 M + U - \eta_k E_c^k) u_c = 0 \tag{4.25}$$

where the matrices E_c^k are the symmetric parts of the originally asymmetric matrices E^k, the skew-symmetric parts being omitted. Since (4.25) is associated with a conservative system, the fundamental characteristic surface defined by

$$|-\omega^2 M + U - \eta_k E_c^k| = 0 \tag{4.26}$$

cannot have convexity toward the fundamental region (Theorem 3.1), and the matrix

$$D_c = -\omega^2 M + U - \eta_k E_c^k \tag{4.27}$$

is positive definite for any point (ω^2, η_k) in its own fundamental region (Section 3.2). This region is shown in Figure 4.2 with shaded boundary. The following lemma can then be stated:

LEMMA 4.1. *The fundamental characteristic surface of the pseudo-conservative system cannot enter the fundamental region of the corresponding conservative system.*

To prove this lemma, it is recalled (Section 1.3, Exercise 7) that if a skew-symmetric matrix is added to a positive-definite matrix, the determinant of the resulting matrix remains positive [44]. This can also readily be verified by associating the two matrices with quadratic forms in certain

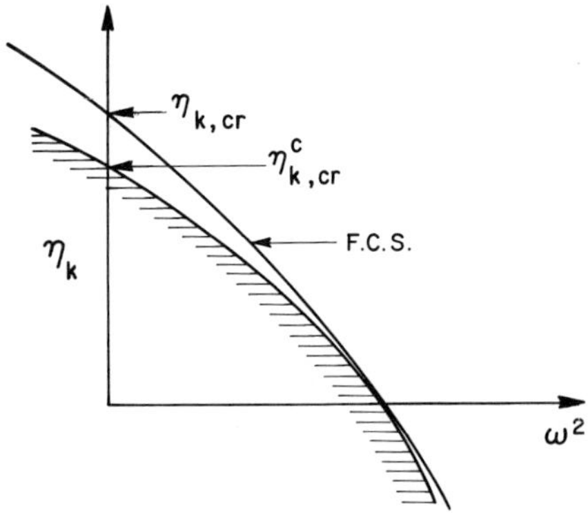

Figure 4.2

arbitrary variables, and observing that the quadratic form associated with the skew-symmetric matrix will always vanish. The proof of the lemma now follows very easily. Suppose the fundamental characteristic surface of the pseudo-conservative system has common points with the fundamental region of the corresponding conservative system; then for such points (ω^2, η_k) the matrix (4.27) is positive definite and

$$|-\omega^2 M + U - \eta_k E^k| > 0 \tag{4.28}$$

since the left-hand side of (4.28) is associated with the sum of a positive definite and a skew-symmetric matrix. Hence, the characteristic surface of the pseudo-conservative system cannot have common points with the fundamental region of the corresponding conservative system since (4.28) should then be an equality. This concludes the proof of the lemma.

It is next noted that for $\eta_k = 0$ the two systems have a common point on the ω^2 axis which is the smallest natural frequency. Keeping this in mind, combining Theorem 3.1, Theorem 4.1 and Lemma 4.1, it is not difficult to deduce the following geometric result:

THEOREM 4.3. *The fundamental characteristic surface of the pseudo-conservative system is tangent to the fundamental characteristic surface of the corresponding conservative system at the ω^2 axis.*

The question arises, of course, whether these two surfaces have only one common point, and it will be discussed shortly. Now, however, another significant result can immediately be reached on the basis of the above theorems and lemma. It is clear from the above results (Fig. 4.2) that the fundamental characteristic surface of the pseudo-conservative system cannot intersect the η_k space below the intersection associated with the corresponding conservative system, and one has:

THEOREM 4.4. *The stability boundary of the corresponding conservative system provides a lower bound (in magnitude) for the stability boundary of the pseudo-conservative system.*

Suppose now that the fundamental characteristic surface of the corresponding conservative system consists of a flat plane(s) (see Theorem 3.3); then by virtue of the above results one formulates the following theorem:

THEOREM 4.5 *If the fundamental characteristic surface of the corresponding conservative system takes the form of a flat plane, the fundamental characteristic surface of the pseudo-conservative system is necessarily the same plane.*

COROLLARY 4.1. *If the fundamental characteristic surface of the corresponding conservative system is a flat plane, the stability boundaries of the two systems coincide and take the form of a flat plane.*

The case in which some η_k represent conservative forces with symmetric E^k deserves further attention. Suppose only one η_k, η_1 say, is conservative; then, the intersection of the plane $\omega^2-\eta_1$ with the fundamental characteristic surface(s) is obviously a single curve for both systems (Fig. 4.3). The two surfaces, however, are distinct outside the $\omega^2-\eta_1$ plane providing they are not flat planes as in Theorem 4.5. In other words, the two surfaces are tangential along the curve associated with the conservative parameter η_1, and this argument holds for any conservative parameters that may be present. Thus, one has:

THEOREM 4.6. *The two fundamental characteristic surfaces have a common point, a line, or an m_1-dimensional surface according to whether*

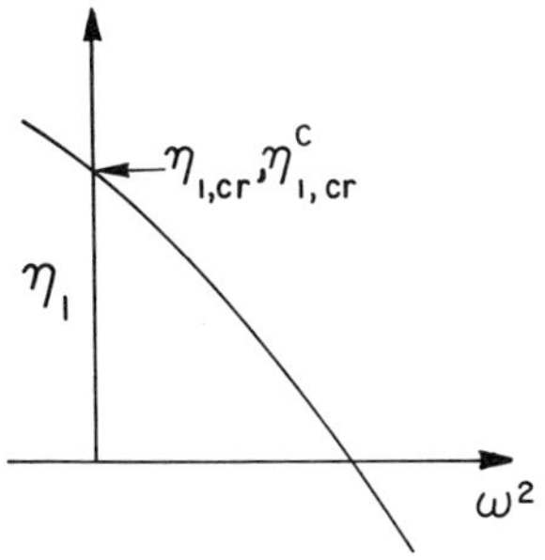

Figure 4.3

the pseudo-conservative system has no, one, or m_1 conservative parameters respectively, and the two surfaces are tangential at these common points.

A similar theorem can be stated for the stability boundaries of the two systems. Clearly, the critical values of the conservative parameters coincide for both systems but the critical values of circulatory parameters are, in general, different. Thus, the following theorem can be stated:

THEOREM 4.7. *The stability boundaries of the two systems have nothing, a point, a curve, or an $(m_1 - 1)$ dimensional surface in common according to whether there exist no, one, two, or m_1 conservative parameters respectively.*

Fig. 4.4(a) illustrates schematically the above phenomena for η_1 and η_2 both conservative, Figure 4.4(b) for η_1 conservative and η_2 circulatory, and Figure 4.4(c) for η_1 and η_2 both circulatory.

The lower bound theorems discussed above can also be established through an appropriate extension of the analysis adopted in Section 4.3 which employs the extremum properties of the Rayleigh quotient. To give an idea on the application of such variational principles in this context,

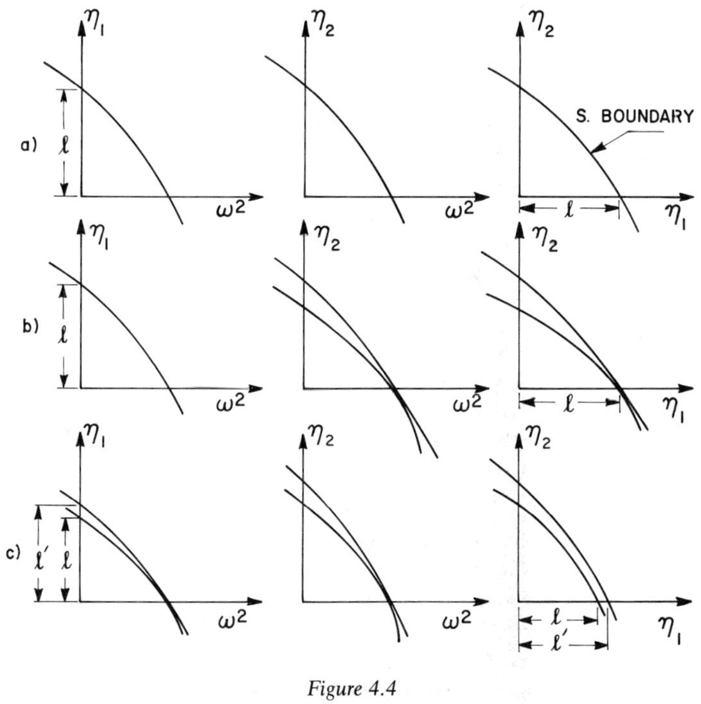

Figure 4.4

suppose the system is a simple one-parameter system such that $\eta_k = l_k\xi$ where the l_k are the direction cosines as before. The pseudo-conservative system is then described by

$$(-\omega^2 M + U - \xi E)u = 0 \tag{4.29}$$

where $E = l_k E^k$, and the corresponding conservative system by

$$(-\omega^2 M + U - \xi E_c)u_c = 0 \tag{4.30}$$

where E_c is the symmetric part of E.

The eigenvalues ω_c^2 of (4.30) at a fixed value of ξ are given by

$$\omega_{ic}^2 = \frac{\langle u_c^i, (U - \xi E_c)u_c^i\rangle}{\langle u_c^i, Mu_c^i\rangle} \tag{4.31}$$

where ω_{1c}^2 will be assumed to be the smallest eigenvalue.

On the other hand the corresponding (at the same ξ) smallest eigenvalue of the pseudo-conservative system can be derived from (4.29) as

$$\omega_1^2 = \frac{\langle u^1, (U - \xi E)u^1\rangle}{\langle u^1, Mu^1\rangle} = \frac{\langle u^1, (U - \xi E_c)u^1\rangle}{\langle u^1, Mu^1\rangle} \tag{4.32}$$

since the skew-symmetric part of E vanishes in the numerator. By comparing (4.31) and (4.32), one obtains

$$\omega_1^2 \geq \omega_{1c}^2 \tag{4.33}$$

which shows that the fundamental frequency of the corresponding conservative system provides a lower bound for the fundamental frequency of the pseudo-conservative system (Fig. 4.5). For $\xi = 0$ one obtains the

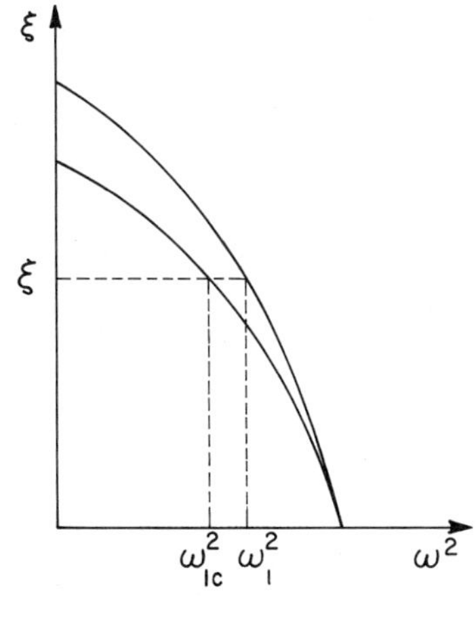

Figure 4.5

natural frequency $\omega_1^2 = \omega_{1c}^2$. As ξ is increased in magnitude, the fundamental frequency of the corresponding conservative system tends to zero at a lower level of ξ due to inequality (4.33). Instability occurs when ω^2 passes through zero to negative values, and therefore the critical parameter ξ (magnitude) of the corresponding conservative system is a lower bound for that of the pseudo-conservative system.

4.5 Practical considerations

Theorems 4.1 and 4.2 can be used to estimate the frequencies of free vibrations at certain values of the parameters η_k and/or to estimate the stability boundary of the system. For this purpose, one of the various approaches described in Section 3.4 may be followed. As a matter of fact, the arguments advanced in Section 3.4 apply directly to pseudo-conservative systems as well and will not, therefore, be repeated here.

The practical implications of the remaining results will now be discussed briefly. Suppose a particular nonconservative system is given, and it is established that the system is not capable of exhibiting flutter at all, thus falling within the scope of this chapter. A lower bound for the stability boundary can then be obtained by using the results of *the corresponding conservative system* and Theorem 4.4. The stability boundary of the latter system may already be given in the literature and such an estimate follows without any analytical effort. Pflüger's rod (see Section 4.7), for example, demonstrates the effectiveness of this approach. The critical value of the tangential follower force of Pflüger's rod is known [42] to be $\eta_{cr} = 18{\cdot}96\, EI/l^3$; the corresponding conservative system with uniformly distributed constant-directional force (instead of tangential follower force), on the other hand, has a critical value of $\eta_{cr}^c = 18{\cdot}58\, EI/l^3$. Clearly, the lower bound estimate is a quite close one, and this example alone gives an indication of the significance of Theorem 4.4 in obtaining a *safe* estimate. It must immediately be remarked here that the conservative parameter η^c of the corresponding conservative system is not necessarily the conservative component of the circulatory parameter, although this happens to be the case in many problems as a result of structural properties. In general, however, the definition given in Section 4.4 must be applied, which stipulates that the corresponding conservative system is obtained by deleting the skew-symmetric parts of the asymmetric matrices E^k.

The lower bound will be the exact solution if Theorem 4.5 is satisfied. Such a situation is produced when the equations of the corresponding conservative system uncouple, resulting in a fundamental characteristic surface which is in the form of a flat plane.

From a dynamics point of view, Theorems 4.3, 4.4 and 4.5 can also be helpful in estimating the frequencies of the pseudo-conservative system on the basis of the corresponding conservative system.

Similar considerations are valid for the remaining results, and their interpretation is left to the reader.

Identification of the system

Clearly, it is desirable to distinguish *a priori* a pseudo-conservative system from other nonconservative systems. An ability to do so will obviously facilitate the application of the general results to obtain estimates concerning the stability properties and dynamic response of a given system without an extensive analysis.

Consider first a two-degree-of-freedom system. Let

$$A = M^{-1}K$$

be symmetrizable. The characteristic equation

$$\begin{vmatrix} a_{11}-\lambda & a_{12} \\ a_{21} & a_{22}-\lambda \end{vmatrix} = 0$$

has the discriminant

$$(a_{11}-a_{22})^2 + 4a_{12}a_{21}$$

which is positive whenever a_{12} and a_{21} have the same sign. It follows that a sufficient condition for the eigenvalues to be real is given by

$$\text{sign } a_{12} = \text{sign } a_{21}. \tag{4.34}$$

Thus, if this condition is satisfied for all values of parameters, the associated system is pseudo-conservative. This simple condition, however, is not necessary, and can of course be replaced by the inequality

$$(a_{11}-a_{22})^2 + 4a_{12}a_{21} > 0$$

which is both necessary and sufficient for the system to have real eigenvalues.

In the general case of n degrees-of-freedom, distinguishing a pseudo-conservative system from other nonconservative systems may not be a simple matter. Symmetric boundary conditions appear to have an influence in producing the distinctive characteristics of this class of systems but cannot be definitive [15]. Mathematically, the system must have real eigenvalues for the entire range of the parameters. Various definitions and properties of symmetrizable matrices given in Section 1.4 may be used to ascertain whether $A = M^{-1}K$ is symmetrizable. According to one of these definitions, a given asymmetric matrix A is symmetrizable if there exist matrices S_1 and S_2 such that

$$S_1^{-1}A = S_2 \tag{4.35a}$$

where S_1^{-1} is a symmetric positive definite matrix and S_2 is symmetric. In many practical situations, S_1^{-1} can be chosen as positive diagonal (see Section 4.7), and the problem is then greatly simplified. Thus, one sets

$$DA = S \tag{4.35b}$$

where D is a positive diagonal matrix with elements $d_{ii} > 0$ and S is symmetric, and ascertains whether or not it is possible to construct the matrix D satisfying the above equation. As it will be demonstrated in the solution of Example 4.7, this approach leads to the desired answer immediately without much effort, and the general results of the preceding sections become available for the system.

If one attempts to generalize this approach by removing the restriction of diagonality on S_1^{-1}, computations might be lengthy and not much easier than determining the eigenvalues directly. Such a more general test can also be performed on the basis of the fact that a symmetrizable matrix is similar to its transpose through

$$A' = S^{-1}AS$$

where S is positive definite (see (1.38)). Then, one has

$$SA' = AS$$

or

$$SA' - AS = 0 \tag{4.36}$$

to solve for a positive definite S. If $S = S' > 0$ exists, the matrix A is symmetrizable.

4.6 Effect of damping

As in the case of conservative systems, if damping forces, derivable from the Rayleigh dissipation function (3.48), are added to the pseudo-conservative system, the equations of motion take the form

$$M\ddot{q} + D\dot{q} + Kq = 0 \tag{4.37}$$

where D is non-negative definite. This equation can, of course, be transformed into a first order equation similar to (3.50). Here, however, the discussion will be based on the second order equation (4.37).

Assuming solutions of the form

$$q = ue^{\lambda t},$$

(4.37) yields

$$(M\lambda^2 + D\lambda + K)u = 0 \tag{4.38}$$

or

$$(\lambda^2 I + M^{-1}D\lambda + M^{-1}K)u = 0 \tag{4.39}$$

where both $M^{-1}K$ and $M^{-1}D$ are symmetrizable and the latter has positive eigenvalues if D is positive definite. Introducing

$$u = T^{-1}y, \qquad T > 0,$$

pre-multiplying (4.39) by T and using (4.10) results in

$$(\lambda^2 I + B\lambda + C)y = 0 \tag{4.40}$$

where $C = T^{-1}S_2T^{-1}$ is symmetric and $B = TM^{-1}DT^{-1}$, which is, in general, asymmetric, has positive eigenvalues if $D > 0$. It follows that if B is expressed as the sum of a symmetric and skew-symmetric matrix; i.e., if

$$B = H + G \quad \text{where} \quad H = H' \quad \text{and} \quad G = -G', \tag{4.41}$$

then the eigenvalue problem (4.40) takes the form

$$[\lambda^2 I + (H + G)\lambda + C]y = 0. \tag{4.42}$$

It is interesting to observe that (4.42) actually describes a damped gyroscopic system which was defined in Section 2.4. In other words, the behaviour of a dissipative pseudo-conservative system appears to be similar to that of a dissipative gyroscopic system with a damping matrix H. If $H > 0$, then the systems described by (4.42) fall within the scope of the Kelvin-Tait-Chetaev (KTC) theorem [25] which provides a useful criterion of stability. According to this theorem, a completely damped gyroscopic system is asymptotically stable (unstable) if the corresponding nondissipative nongyroscopic system, which is characterized here by $H = G = 0$, is stable (unstable). The KTC theorem was originally stated by Kelvin and Tait and later proved by Chetaev. An alternative proof will be given in Section 5.10, and only its interpretation in the present context will be discussed here.

It follows from the KTC theorem that the system (4.42) is asymptotically stable if C and H are positive definite. An important point is now noted. Although B has real positive eigenvalues if $D > 0$, it does not necessarily follow that in this case H is positive definite. In fact, it is not difficult to verify in general that the positive definiteness of the symmetric part (e.g. H) of a symmetrizable matrix (e.g. B) is sufficient but not necessary for the latter to have positive eigenvalues. Alternatively, it is necessary but not sufficient that a given symmetrizable matrix has positive eigenvalues for its symmetric part to be positive definite.

This observation has serious implications with regard to the stability properties of the system (4.42). It is clear from the above remarks that even if $D>0$, H can be indefinite. On the other hand, it can be shown (Section (5.10)) that for the stability (asymptotic) of a gyroscopic system of the form (4.42) the conditions

$$h>0, \qquad c>0$$

must be satisfied for all eigenvectors; where h and c are obtained by pre-multiplying (4.42) by $\bar{y}'$ (complex conjugate), such that $\langle \bar{y}, y \rangle = 1$, $\langle \bar{y}, Hy \rangle = h$, $\langle \bar{y}, Cy \rangle) = c$. Since H can be indefinite, however, the condition $h>0$ may not be satisfied even if $c>0$.

It then follows that damping may have a destabilizing effect on the behaviour of a pseudo-conservative system.

It is remarkable that while the behaviour of an undamped pseudo-conservative system is completely analogous to that of a conservative system, as has been demonstrated in the preceding sections, the addition of dissipative forces may alter this behaviour significantly, exposing the nonconservative characteristics of the system. It is of-course understood that the analogy with conservative systems may continue to exist. In the special case of $G=0$, for example, $H>0$ if $D>0$, and the system (4.42) is asymptotically stable provided $C>0$. An example of this situation is provided by Pflüger's column which will be studied in the following section. In this example the matrices M and T are diagonal and if one adds viscous damping terms represented by a diagonal matrix $D>0$, then the matrix B is also diagonal and positive definite, indicating asymptotic stability provided $C>0$. The system loses stability by *divergence* when C ceases to be positive definite.

The case $D \geq 0$ and pervasive damping will further be discussed in Section 5.10 (see also Section 5.8) where more general velocity-dependent forces are also considered.

4.7 Example: Pflüger's column

In analogy with Example 2 of Chapter 3, consider now the simply-supported elastic bar subjected to uniformly distributed *tangential* forces

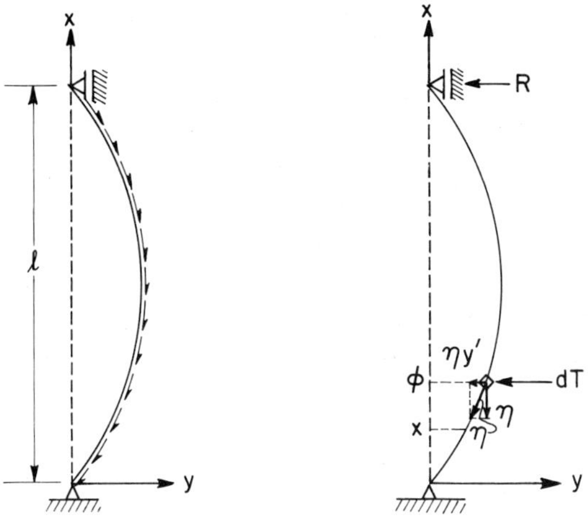

Figure 4.6

rather than constant-directional forces (Fig. 4.6). To obtain the governing differential equation of small free vibrations, the inertia forces

$$dT = m\ddot{y}(\varphi, t)\, d\varphi \tag{4.43}$$

must be taken into account. Writing the bending moment at x, one has

$$sy''(x, t) = -R(l-x) + \int_x^l \eta[y(\varphi, t) - y(x, t)]\, d\varphi - \int_x^l \eta(\varphi - x)y'(\varphi, t)\, d\varphi - m\int_x^l (\varphi - x)\ddot{y}(\varphi, t)\, d\varphi \tag{4.44}$$

where dots denote differentiation with respect to time t, s is the constant flexural rigidity (EI is not used to avoid confusion with the matrix E) and the boundary conditions are

$$\begin{aligned} y(0, t) &= y(l, t) = 0, \\ y''(0, t) &= y''(l, t) = 0. \end{aligned} \tag{4.45}$$

Differentiating (4.44) with respect to x results in

$$sy'''(x, t) = R - \int_x^l \eta y'(x, t)\, d\varphi + \int_x^l \eta y'(\varphi, t)\, d\varphi + m\int_x^l \ddot{y}(\varphi, t)\, d\varphi, \tag{4.46}$$

and a second differentiation yields

$$sy''''(x, t) = -\eta(l-x)y''(x, t) + \eta y'(x, t) - \eta y'(x, t) - m\ddot{y}(x, t). \tag{4.47}$$

Hence, the differential equation is in the form

$$sy'''' + \eta(l-x)y'' + m\ddot{y} = 0. \tag{4.48}$$

Before proceeding further, note that the differential equation governing the motion of the conservative system with uniformly distributed constant-directional forces (Example 2, Section 3.7) is immediately available from (4.47) as

$$sy'''' + \eta(l-x)y'' - \eta y' + m\ddot{y} = 0. \tag{4.49}$$

It will be seen later that this equation actually represents "the corresponding conservative system" defined in Section 4.4.

Returning now to the nonconservative system under consideration, suppose the function $y(x, t)$ is in the form

$$y(x, t) = e^{i\omega t} z(x). \tag{4.50}$$

Introducing (4.50) into (4.48) then yields

$$sz'''' + \eta(l-x)z'' - \omega^2 mz = 0 \tag{4.51}$$

with the boundary conditions

$$\begin{aligned} z(0) &= z(l) = 0, \\ z''(0) &= z''(l) = 0. \end{aligned} \tag{4.52}$$

For the solution of (4.51) assume a deflection function of the form (3.81) and apply the Galerkin procedure. Thus, introducing

$$\begin{aligned} z &= u_1 \sin\frac{\pi x}{l} + u_2 \sin\frac{2\pi x}{l} \\ &\equiv u_1 z_1(x) + u_2 z_2(x) \end{aligned} \tag{4.53}$$

into (4.51), multiplying by z_1 and z_2, and integrating over x results in

$$(-M\omega^2 + U - \eta E)u = 0 \tag{4.54}$$

where

$$M = m\begin{bmatrix} \int_0^l z_1^2 \, dx & \int_0^l z_1 z_2 \, dx \\ \int_0^l z_2 z_1 \, dx & \int_0^l z_2^2 \, dx \end{bmatrix} = \frac{lm}{2}\begin{bmatrix} 1 & 0 \\ 0 & 1 \end{bmatrix}, \tag{4.55}$$

$$U=\begin{bmatrix}\int_0^l sz_1''''z_1\,\mathrm{d}x & \int_0^l sz_2''''z_1\,\mathrm{d}x\\ \int_0^l sz_1''''z_2\,\mathrm{d}x & \int_0^l sz_2''''z_2\,\mathrm{d}x\end{bmatrix}=\begin{bmatrix}s\dfrac{\pi^4}{2l^3} & 0\\ 0 & s\dfrac{8\pi^4}{l^3}\end{bmatrix}, \tag{4.56}$$

$$E=\begin{bmatrix}\int_0^l (l-x)z_1''z_1\,\mathrm{d}x & \int_0^l (l-x)z_2''z_1\,\mathrm{d}x\\ \int_0^l (l-x)z_1''z_2\,\mathrm{d}x & \int_0^l (l-x)z_2''z_2\,\mathrm{d}x\end{bmatrix}=\begin{bmatrix}\dfrac{\pi^2}{4} & \dfrac{32}{9}\\ \dfrac{8}{9} & \pi^2\end{bmatrix}, \tag{4.57}$$

and $u=(u_1, u_2)'$.

Why is the system described by (4.54) pseudo-conservative? According to the theory, if the system is pseudo-conservative, then the matrix $M^{-1}(U-\eta E)$ must be symmetrizable for all values of η. Recalling the fact that if an asymmetric matrix is symmetrizable, there must exist a symmetric positive definite matrix such that the relation (4.35a) is satisfied, one looks for such a matrix. For simplicity, it is first attempted to satisfy (4.35b) as explained in Section 4.5 with a positive definite diagonal matrix (this is a sufficient condition). Let

$$D=d_i\,\delta_{ij} \tag{4.58}$$

so that

$$DM^{-1}(U-\eta E)=S \tag{4.59}$$

where S is symmetric. This condition is satisfied in this example by choosing, for instance,

$$D=\begin{bmatrix}1 & 0\\ 0 & 4\end{bmatrix} \tag{4.60}$$

and hence the system is a pseudo-conservative one. From (4.59) one has

$$S=\frac{1}{m}\begin{bmatrix}s\dfrac{\pi^4}{l^4}-\eta\dfrac{\pi^2}{2l} & -\eta\dfrac{64}{9l}\\ -\eta\dfrac{64}{9l} & s\dfrac{64\pi^4}{l^4}-\eta\dfrac{8\pi^2}{l}\end{bmatrix} \tag{4.61}$$

and the eigenvalue problem (4.54) can be written as

$$(-\omega^2D+S)u=0 \tag{4.62}$$

which corresponds to (4.16a) with $D=T^2$ and $S=S_2$. Alternatively, (4.62) can be expressed as

$$(-\omega^2I+C)y=0 \tag{4.63}$$

where

$$C = D^{-\frac{1}{2}}SD^{-\frac{1}{2}} \quad \text{and} \quad u = D^{-\frac{1}{2}}y.$$

The *corresponding conservative system* as defined in Section 4.4 can readily be obtained by deleting the skew-symmetric part of the asymmetric matrix E given in (4.57). This operation yields

$$(-\omega^2 M + U - \eta E_c)u_c = 0 \tag{4.64}$$

where

$$E_c = \begin{bmatrix} \dfrac{\pi^2}{4} & \dfrac{20}{9} \\ \dfrac{20}{9} & \pi^2 \end{bmatrix}. \tag{4.65}$$

As expected, the corresponding conservative system (4.64) is, indeed, the dynamical description of the system discussed in Section 3.8 and characterized by the eigenvalue problem (3.83). The critical load for this system was found to be $\eta^c_{cr} = 18{\cdot}58\, s/l^3$. The critical value of η for the pseudo-conservative system (4.54) can be found by setting $\omega^2 = 0$ in the characteristic equation. Thus,

$$|U - \eta E| = 0 \tag{4.66}$$

yields

$$\eta_{cr} = 18{\cdot}96\, \frac{s}{l^3} \tag{4.67}$$

which indicates that η^c_{cr} is a good lower bound to (4.67).

The Rayleigh quotient can be formed in two ways:

$$R(x) = \frac{\langle x, Sx\rangle}{\langle x, Dx\rangle}$$

which follows from (4.62), and

$$R(x) = \frac{\langle x, Cx\rangle}{\langle x, x\rangle}$$

from (4.63).

The reader is now in a position to demonstrate the main results of the theory presented in this chapter by means of this example and others that can readily be formulated on the basis of the foregoing analysis. The following problems are designed for this purpose.

Problems

1. Show that the smallest eigenvalue ω^2 of the system (4.54) is larger than that of the system (4.64), and demonstrate the convexity of the fundamental characteristic curve.

2. Analyze the system (4.51) by introducing a deflection function in the series form

$$z = \sum_{k=1}^{n} u_k \sin \frac{k\pi x}{l}$$

and show that the resulting n-degree-of-freedom system is pseudo-conservative.

3. Consider the generalized Pflüger problem [49, 50] in which an elastic column with simply supported ends is subjected to a distributed tangential load $\eta f(x)$ where

$$f(x) = \int_0^x g(\varphi)\, \mathrm{d}\varphi.$$

Analyze this system similarly by using the sine series and show that it is a pseudo-conservative system.

If the loading consists of a single concentrated tangential load at $x = l/2$, one can write $g(x) = \delta(x - l/2)$. Show that this system has a critical load of $\eta_{\mathrm{cr}} = 16{\cdot}57 s/l^2$ as compared to $\eta_{\mathrm{cr}}^{\mathrm{c}} = 15{\cdot}49 s/l$ of the corresponding conservative system.

4. Analyze Pflüger's column when in addition to tangential forces a concentrated axial compression force (conservative) acts simultaneously. Show that the corresponding conservative system is the one described in Problem 1 of Section 3.7. Demonstrate all the theorems established in the theory.

5

Gyroscopic systems

A remarkable class of linear dynamical systems is concerned with gyroscopic forces, and gyroscopic systems were defined in Section 2.4 where (2.64) was derived as the governing differential equation of an internally and externally damped gyroscopic system. The actual process leading to (2.64) in practical applications will be demonstrated when examples are formulated (see Example 4). An important special case of (2.64) is concerned with undamped systems which were described in Section 2.4 as *gyroscopic conservative.* These systems will be treated first in the following sections before the more general gyroscopic systems are considered.

If the structure of the matrices in (2.64) is examined carefully, one notes that this equation can describe a variety of systems as well as gyroscopic systems when a broader interpretation is adopted. In fact, the conditions of stability and instability pertaining to linear, holonomic, autonomous mechanical systems discussed and reviewed by several authors in general terms are based on matrix equations similar to (2.64). Magnus [51], for example, gave a classification of mechanical systems with respect to various types of forces on the basis of the properties of the matrices involved. Earlier, Bottema [52] considered a two-degree-of-freedom general system and explored various possible types of behaviour under certain loading conditions. A comprehensive paper by Shieh and Masur [53], originally intended for elastic and Voigt-Kelvin solid bodies, contains mechanical systems described by an equation similar to (2.64) as a special case and produces general stability criteria. In a more recent (1975) review of vibrations and stability of mechanical systems [54], the discussion is based on a differential equation analogous to (2.64). It should, therefore, be kept in mind that the developments in this chapter may have significant applications in other contexts, although the discussion here will be oriented toward gyroscopic systems.

5.1 Gyroscopic conservative systems

Vibrations and stability of gyroscopic conservative systems have been studied by several authors following Thompson and Tait [55]. Ziegler's book [13] contains a general discussion, and one can find a variety of investigations ranging from important practical problems to new general theories in the proceedings [56] of a recent IUTAM symposium (1974, Lyngby, Denmark) edited by Niordson. General developments include [57–63].

Consider the dynamical autonomous system described by

$$M\ddot{q} + G\dot{q} + Kq = 0 \tag{5.1}$$

where $M = M' > 0$, $K = K'$ and $G = -G'$. This equation follows from Lagrange's equations of motion if the kinetic and potential energies are assumed to be in quadratic forms as usual. Clearly, besides the conservative forces derivable from a potential the only velocity-dependent forces are of gyroscopic type as explained in Section 2.4.

It can readily be shown that the systems described by (5.1) are conservative. Thus, considering the Hamiltonian

$$H = \tfrac{1}{2}\langle \dot{q}, M\dot{q}\rangle + \tfrac{1}{2}\langle q, Kq\rangle$$

one observes that

$$\dot{H} = -\langle \dot{q}, G\dot{q}\rangle = 0$$

which shows that the energy is conserved during motion. It also follows that if the potential energy, and hence the matrix K, is positive definite, then the Hamiltonian is a Lyapunov function and the system is stable. This is, however, a sufficient condition only, and the positive definiteness of the potential energy ceases to be a necessary condition for stability in contrast with non-gyroscopic conservative systems. In other words, the gyroscopic forces may stabilize a conservative system which would have been unstable in their absence. This remarkable effect of gyroscopic forces and the process leading to it will be illustrated clearly by the properties of the characteristic curves when the explicit equations of motion are considered. At this stage, however, one can also show that the system may be stable even when K is not positive definite.

Assuming solutions in the form

$$q = u e^{\lambda t} \tag{5.2}$$

gives rise to the matrix equation

$$(\lambda^2 M + \lambda G + K)u = 0. \tag{5.3}$$

Consider first the case in which K is positive semi-definite. Then, pre-multiplying (5.3) by $\bar{u}$, one obtains

$$\lambda^2 m + i\lambda g + k = 0 \tag{5.4}$$

where $m = \langle \bar{u}, Mu\rangle > 0$, $ig = \langle \bar{u}, Gu\rangle$, $k = \langle \bar{u}, Ku\rangle \geq 0$ and g is real.

The solution can be expressed as

$$\lambda = i\left(\frac{-g \pm \sqrt{g^2+4mk}}{2m}\right) \tag{5.5}$$

which indicates that all the eigenvalues are imaginary as long as $k>0$. For $k=0$, one of the corresponding eigenvalues vanishes and the system exhibits a critical divergence point. Clearly, $k<0$ may result in a complex eigenvalue with a positive real part, indicating instability; the condition $k<0$, however, is not sufficient (although necessary) for instability, and if the inequality

$$g^2+4mk>0 \tag{5.6}$$

holds for all eigenvectors the system is stable even if K is not positive definite and some k is negative. Evidently, one can obtain general stability criteria concerning the system (5.3) on the basis of (5.5) conveniently. For this purpose, let K be a symmetric matrix (not necessarily definite) and

$$\lambda = \gamma + i\omega \tag{5.7}$$

where γ and ω are real. The following results can be stated:

(1) The system described by (5.3) is stable if and only if the inequality (5.6) holds and $k \neq 0$ for all eigenvectors.

In this case all $\gamma=0$ and no ω is repeated.

(2) The system (5.3) is in a critical state ($\gamma=0$, ω repeated) if either

$$g^2+4mk=0 \qquad \text{(flutter)} \tag{5.8}$$

or

$$k=0 \qquad \text{(divergence)} \tag{5.9}$$

for some eigenvector.

In the latter case, if the remaining eigenvectors satisfy the inequality (5.6) the critical point is primary. A flutter point, on the other hand, can never be primary in the sense that when it occurs the system must have already exhibited a divergence point since (5.8) cannot hold if K is positive definite. It is also noted that at the flutter point the frequency is given by $\omega = -g/2m$ and at the divergence point $g=0$ as well as $\omega=0$ since the corresponding eigenvector is real. Furthermore, the critical points of the linear system are, in fact, unstable.

(3) The system (5.3) is unstable if and only if either

$$g^2+4mk<0, \qquad g \neq 0 \qquad \text{(flutter instability)} \tag{5.10}$$

or

$$k<0, \qquad g=0 \qquad \text{(divergence instability)} \tag{5.11}$$

for at least one eigenvector.

Clearly, when (5.10) holds $\omega=-g/2m\neq 0$ while (5.11) implies that $\omega=0$.

These results follow from (5.5) very easily, but can also be deduced from (5.4) which, upon introduction of (5.7), yields

$$2\gamma\omega m+\gamma g=0 \tag{5.12}$$

and

$$(\gamma^2-\omega^2)m-\omega g+k=0. \tag{5.13}$$

(5.12), for example, clearly shows that either $\gamma=0$ or $\omega=-g/2m$, and the use of these in (5.13) leads to the criteria above. Analogous criteria were presented by Shieh [58] for continuous systems.

In summary, the following important observations are made:

(i) The energy criterion is only a sufficient condition for stability of gyroscopic conservative system.
(ii) While gyroscopic forces can stabilize an otherwise unstable conservative system, they can never destabilize a stable conservative system.
(iii) It is clear from (5.10) that a gyroscopic conservative system can lose stability by flutter although this was shown to be impossible for a conservative system. The significance of flutter instability lies in the fact that it may occur after the gyroscopic forces have stabilized an otherwise unstable conservative system, effectively creating a flutter region between divergence points on the parameter axis. This feature can best be explained by examining the characteristic curves of the system. Before proceeding to do so, however, a sufficient condition for instability in terms of the matrices M, K and G will be derived.

It was shown by Hagedorn [63] that the gyroscopic system (5.1) is unstable if the matrix

$$4K-GM^{-1}G \tag{5.14}$$

is negative definite. Consider the complex vectors $M^{-\frac{1}{2}}Gu$ and $M^{+\frac{1}{2}}\bar{u}$; according to the well-known inequality (Exercise 2, Section 1.2) of scalar products of vectors one has

$$|\langle M^{\frac{1}{2}}\bar{u}, M^{-\frac{1}{2}}Gu\rangle|^2\leq-\langle\bar{u}, Mu\rangle\,.\,\langle\bar{u}, GM^{-1}Gu\rangle \tag{5.15}$$

or

$$g^2+m\langle\bar{u}, GM^{-1}Gu\rangle\leq 0. \tag{5.16}$$

But, if (5.14) is negative definite, then

$$4k < \langle \bar{u}, GM^{-1}Gu \rangle$$

which, upon introduction into (5.16), leads to

$$g^2 + 4mk < 0$$

and hence instability according to (5.10). A generalization of this result can be found in Ref. [64]. Further similar sufficient stability and instability conditions have been given by Walker [65]. For instance, if there exists a symmetric positive definite matrix S such that $SM^{-1}G$ is positive definite and $SM^{-1}K$ is symmetric positive definite, then the system is asymptotically stable. The truth of this result can be inferred from

$$S\ddot{q} + SM^{-1}G\dot{q} + SM^{-1}Kq = 0 \tag{5.17}$$

which under the conditions imposed on S falls within the scope of the *KTC* theorem to be proved in Section 5.10. Similarly, conditions for stability and instability can be established. The condition of positive definiteness of the matrix $SM^{-1}G$ actually can be relaxed. If the symmetric part B of $SM^{-1}G$ is only positive semi-definite, the system is still asymptotically stable if the eigenvectors of the system (5.17) with $B = 0$ do not lie in the null space of B. This is related to the concept of *pervasive damping* and will be discussed further in Section 5.10.

5.2 Characteristic curves

It will be more convenient for the development of this section to assume the solutions in the form of $q = ue^{i\omega t}$ rather than (5.2), and to work with

$$H(\omega)u \equiv (-\omega^2 M + i\omega G + K)u = 0. \tag{5.18}$$

The roots of the characteristic equation

$$|-\omega^2 M + i\omega G + K| = 0 \tag{5.19}$$

are in general complex, but due to the symmetry of M and K and the skew-symmetry of G they appear in pairs ω_k, $-\omega_k$, $k = 1, 2, \ldots, n$. Note that the transpose of the matrix $H(\omega)$ is $H(-\omega)$, and the determinant of a matrix is the same as that of its transpose. It follows that the characteristic polynomial resulting from the expansion of the determinant (5.19) will be a function of ω^2, and this property will form the basis of the following analysis.

It is first noted that the system remains stable as long as all ω_k^2 are real and positive. Furthermore, the divergence points on the parameter axis, corresponding to the vanishing of the ω_k, are the same for both the gyroscopic system and *the corresponding nongyroscopic system* which is defined here as the system described by (5.18) with $G=0$. At this stage, the concept of the parameter mentioned above should be clarified in the context of the present discussion.

Gyroscopic forces in dynamical systems generally are characterized by a parameter ξ, such as a velocity or angular velocity, and the gyroscopic matrix G is proportional to ξ. The square of ξ appears in the matrix K, such that

$$K=U-\xi^2 E, \qquad G=\xi F \tag{5.20}$$

where $U=U'>0$ is associated with the strain energy as before, $E=E'$ and $F=-F'$. ξ^2 will be referred to as the parameter and this section will be concerned with one-parameter systems only, the introduction of additional conservative forces being deferred until Section 5.4. Without loss of generality, the matrices M and U can further be assumed to be the identity matrix I and a diagonal matrix, respectively, since such a transformation can always be achieved on the basis of the positive definiteness of M and U. Then, (5.18) becomes

$$(-\omega^2 I+i\omega\xi F+U-\xi^2 E)u=0 \tag{5.21}$$

where the notation has been kept the same and U is diagonal.

Systems falling within the scope of (5.21) may be divided [59] into three categories:

(a) $E=I$. Rotating shafts when treated in a rotating reference frame attached to the shaft are an example of this case. Then ξ assumes the role of angular velocity and the loading parameter ξ^2 represents the centrifugal force. Other conservative forces are not accounted for in the explicit form (5.21).

(b) E is a diagonal matrix. An example is given by the transverse vibrations of a simply-supported pipe with internal fluid flow of velocity ξ, with a modal analysis using the modes of free vibration.

(c) E is not diagonal. The example in (b) would fit into this category if one or both ends of the pipe were clamped rather than simply supported. Here, when M is transformed to I and U to a diagonal matrix, E does not simultaneously become diagonal.

Clearly, the characteristic polynomial associated with (5.21) is a function of ξ^2 and ω^2. It is, therefore, natural to examine the relationship between ξ^2 and ω^2, rather than working with ξ and ω. Hence, the characteristic curves in the $\omega^2-\xi^2$ plane will be investigated. Since $\xi^2<0$ has no physical meaning for the systems under consideration, it will be assumed that $\xi^2\geq 0$. Points in the first quadrant, where $\omega^2>0$, will be stable, while the region $\omega^2\leq 0$ corresponds to instability (the critical divergent point is also included in the region of instability).

It was stated earlier that the critical divergence values of the loading parameter ξ^2 are unaltered when $G=0$. Similarly, the natural frequencies for $\xi^2=0$ are the same for both the gyroscopic system and the corresponding nongyroscopic system. The distinction between the two systems arises when the intervals between such points are considered. For the corresponding nongyroscopic system, it has been shown (Section 3.2) that the fundamental characteristic curve is concave toward the origin (if not a straight line). A typical case is depicted in (Fig. 5.1), in which the fundamental curve passes through the smallest frequency value on the ω^2 axis, approaches the loading axis with gradually increasing ξ^2 and eventually intersects this axis at the first divergence point ξ_1^2. The fundamental curve of the gyroscopic system also passes through the same points on the axes. As for the behaviour in-between, it will now be proved that it always lies to the left of the fundamental curve of the corresponding nongyroscopic system.

Consider the general system (5.18) and the ω-matrix $H(\omega)$. For any real ω the matrix H is Hermitian. If K is positive definite, the system is stable and the roots ω_k^2 of the characteristic equation (5.19) may be

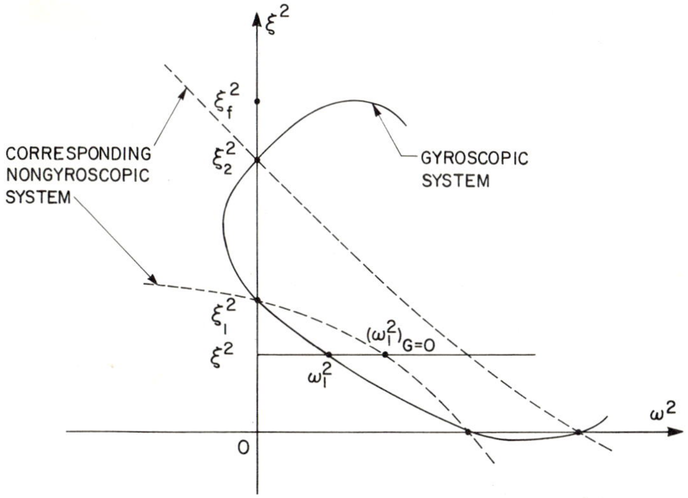

Figure 5.1

ordered in the form

$$0<\omega_1^2\leq\omega_2^2\leq\cdots\leq\omega_n^2. \tag{5.22}$$

Similarly, for the corresponding nongyroscopic system ($G=0$) one can write the roots in the form

$$0<(\omega_1^2)_{G=0}\leq(\omega_2^2)_{G=0}\leq\cdots\leq(\omega_n^2)_{G=0}. \tag{5.23}$$

Now consider the following lemma [59]:

LEMMA 5.1. *If the matrix K in (5.18) is positive definite, then the Hermitian matrix $H(\omega)$ is positive definite for any real ω satisfying $0\leq\omega^2<\omega_1^2$.*

PROOF. Let the eigenvalues of $H(\omega)$ be $\lambda_k(\omega)$, $k=1,2,\ldots,n$. Since $H(0)=K$ is positive definite, all $\lambda_k(0)$ are positive and $\det|H(0)|$ is positive. In addition, since $A+\varepsilon B$ is positive definite for arbitrary Hermitian matrices A and B with A positive definite and ε sufficiently small (Exercise 6, Section 1.3), it follows that $K-\omega^2M$ and $H(\omega)$ are positive definite for sufficiently small values of ω. Then, using (5.22) and the fact that $\det|H(\omega)|$ only vanishes at $\omega^2=\omega_1^2,\ldots,\omega_n^2$, one has

$$\det|H(\omega)|>0 \quad \text{for} \quad 0\leq\omega^2<\omega_1^2. \tag{5.24}$$

Since

$$\det|H(\omega)|=\lambda_1(\omega)\,.\,\lambda_2(\omega),\ldots,\lambda_n(\omega), \tag{5.25}$$

this product of eigenvalues is also positive for $0\leq\omega^2<\omega_1^2$. Each eigenvalue $\lambda_k(\omega)$ is positive at $\omega=0$ and varies continuously with ω. From (5.25) it is seen that $\lambda_k(\omega)=0$ for any k would imply $\det|H(\omega)|=0$. Therefore, it follows that $\lambda_k(\omega)>0$ for all k if $0\leq\omega^2<\omega_1^2$, and hence $H(\omega)$ is positive definite for $0\leq\omega^2<\omega_1^2$.

The following theorem can then be proved [59]:

THEOREM 5.1. *If the matrix K is positive definite, then*

$$0<\omega_1^2\leq(\omega_1^2)_{G=0}. \tag{5.26}$$

PROOF. For $0\leq\omega^2<\omega_1^2$ it follows from the lemma above that

$$\langle\bar{x},H(\omega)x\rangle>0 \tag{5.27}$$

where x is an arbitrary complex nonzero n-vector. Now consider real vectors x, so that $x=\bar{x}$; since G is skew-symmetric,

$$\langle x,H(\omega)x\rangle=\langle x,(K-\omega^2M)x\rangle \tag{5.28}$$

which is positive due to (5.27), and hence the symmetric matrix $K-\omega^2 M$ is positive definite for $0\leq\omega^2<\omega_1^2$. The smallest root $(\omega_1^2)_{G=0}$ of the equation $\det|K-\omega^2 M|=0$ therefore satisfies (5.26) and the theorem is proved.

Clearly, $K=U-\xi^2 E$ is positive definite for $0\leq\xi^2<\xi_1^2$ and the theorem furnishes an upper bound for the fundamental frequencies of the gyroscopic system in the first stability interval. As depicted in Figure 5.1, the fundamental characteristic curve of the system (5.21) lies to the left of the one for the corresponding nongyroscopic system when $0\leq\xi^2<\xi_1^2$. In other words, no characteristic curve of the $G=0$ system can enter into the region in the first quadrant between the origin and the fundamental curve of the gyroscopic system.

Before continuing this discussion, an alternative interesting proof of Theorem 5.1 will be presented [60].

PROOF 2. It was seen that the matrix $H(\omega)$ is positive definite when $\omega=0$, and its smallest eigenvalue λ_1 vanishes for $\omega=\omega_1$ which leads to

$$\langle\bar{u}^1, H(\omega_1)u^1\rangle=0 \tag{5.29}$$

where u^1 is the corresponding eigenvector. Using the extremum properties of the Rayleigh quotient for Hermitian matrices then yields

$$\langle v^1, H(\omega_1)v^1\rangle\geq 0 \tag{5.30}$$

where the real vector v^1 corresponds to $(\omega_1^2)_{G=0}$ of the nongyroscopic system

$$(-\omega^2 M+K)v=0 \tag{5.31}$$

and is assumed to be normalized with respect to M such that

$$\langle v^1, Mv^1\rangle=1. \tag{5.32}$$

It then follows from (5.30) that

$$-\omega_1^2+(\omega_1^2)_{G=0}\geq 0 \tag{5.33}$$

since

$$\langle v^1, Gv^1\rangle=0 \quad\text{and}\quad (\omega_1^2)_{G=0}\langle v^1, Kv^1\rangle. \tag{5.34}$$

In cases (a) and (b) and in case (c) if U and E commute (so that U and E could have been diagonalized simultaneously), the fundamental curves of the system (5.31) are straight lines in the $\omega^2-\xi^2$ plane. The dashed lines in Figure 5.1 would then be straight and the upper bound on ω_1^2 would have a particularly simple form. Also, if the fundamental curve for the $G=0$ system is a straight line and the fundamental curve for the gyroscopic system is a second-order curve in the $\omega^2-\xi^2$ plane (e.g., if $n=2$), it follows that the portion of this latter curve in the region where $K>0$ is *convex* toward the origin.

As ξ^2 is increased beyond ξ_1^2, the fundamental curve of the gyroscopic system may remain in the $\omega^2>0$ region passing the point ξ_1^2 tangentially or it may enter the instability region $\omega^2<0$. In the latter case it has to stay above the fundamental curve of the $G=0$ system. This is due to the fact that for $\omega^2<0$ the matrix $H(\omega)$ is real and, with $\omega=i\gamma$, it has the form

$$H(i\gamma)=\gamma^2M+\gamma G+K \tag{5.35}$$

whose determinant cannot vanish when $K+\gamma^2M$ is positive definite (Exercise 7, Section 1.3). The fundamental curve of the gyroscopic system normally reenters the stable quadrant $\omega^2>0$ at the second divergent point ξ_2^2 on the load axis (Fig. 5.1). One can then interpret this behaviour as the *stabilization of a conservative system by gyroscopic forces*. The higher order characteristic curves behave similarly and a succession of stability and instability regions are produced. The instability region need not be associated with divergence as explained in the following section.

5.3 Flutter instability

For $\xi^2<\xi_1^2$ all the values ω_k^2 are necessarily real; however, there is no guarantee that they will remain real for $\xi^2>\xi_1^2$. In fact, it is clear from the instability condition (5.10) that these roots can be complex. Following the characteristic curve, complex values can occur when it reaches a peak with respect to ξ^2 just as for non-dissipative circulatory systems. As depicted in Figure 5.1, at $\xi^2=\xi_f^2$, a curve may have a maximum point at which

$$\frac{d(\xi^2)}{d(\omega^2)}=0 \tag{5.36}$$

and above which two roots ω_k^2 become complex conjugate, resulting in unstable oscillations.

It is noted that the gyroscopic system described by (5.18) is, in general, nonself-adjoint, but it becomes self-adjoint for real $\omega^2>0$. Thus one has

$$v'(-\omega^2M+i\omega G+K)=0 \tag{5.37}$$

associated with (5.18), and for real $\omega^2>0$, $v=\bar{u}$, while for real $\omega^2<0$, u

and v are real. In other words, for real $\omega^2 > 0$ the left eigenvectors of the system are given by complex conjugates of the right eigenvectors.

Post-multiplying (5.37) by u yields

$$\langle v, (-\omega^2 M + i\omega G + K)u \rangle = 0. \tag{5.38}$$

The left side of (5.38) is stationary with respect to changes in u and v satisfying (5.18) and (5.37) respectively.

First consider the region of real $\omega^2 > 0$, in which $v = \bar{u}$. Define the quantities

$$m = \langle \bar{u}, Mu \rangle, \qquad ig = \langle \bar{u}, Gu \rangle, \qquad k = \langle \bar{u}, Ku \rangle \tag{5.39}$$

as before so that (5.38) becomes

$$-\omega^2 m - \omega g + k = 0. \tag{5.40}$$

Solving for ω yields the flutter condition

$$g^2 + 4mk = 0 \tag{5.41}$$

as in (5.8). In order to obtain the value of the loading parameter at the onset of flutter, pre-multiply (5.21) by v' to get

$$\langle v, (-\omega^2 I + i\omega\xi F + U - \xi^2 E)u \rangle = 0. \tag{5.42}$$

Differentiating (5.42) with respect to ω and using (5.36) results in the flutter condition

$$2\omega \langle v, u \rangle = i\xi \langle v, Fu \rangle. \tag{5.43}$$

Eliminating ω between (5.42) and (5.43) yields the flutter load [59]

$$\xi^2 = a\left[e - \frac{f^2}{4b}\right]^{-1} \tag{5.44}$$

where

$$a = \langle \bar{u}, Uu \rangle, \qquad e = \langle \bar{u}, Eu \rangle, \quad \text{if} = \langle \bar{u}, Fu \rangle, \qquad b = \langle \bar{u}, u \rangle \tag{5.45}$$

in the region of real $\omega^2 > 0$.

If flutter occurs in the region where $\omega^2 < 0$, the flutter load can similarly be determined in terms of left and right eigenvectors as

$$\xi_f^2 = a\left[e + \frac{f^2}{4b}\right]^{-1} \tag{5.46}$$

where

$$a = \langle v, Uu \rangle, \qquad f = \langle v, Fu \rangle, \qquad e = \langle v, Eu \rangle \quad \text{and} \quad b = \langle v, u \rangle.$$

Considering the first quadrant, an interesting distinction between the flutter of nondissipative circulatory systems described by (2.55), which will be studied in Chapter 6, and gyroscopic conservative systems is observed: while the left and right cigenvectors shift from the real to the complex domain (not being complex conjugate) at the onset of flutter in the former case, they are complex immediately before and after flutter in the latter case and lose their complex conjugate status at flutter.

Finally, it is emphasized that since flutter instability may enter the picture after the gyroscopic forces have stabilized the equilibrium, if one limits the investigation to divergence instability (the system being essentially conservative) the loss of stability due to flutter in an otherwise stable region would be overlooked. On the other hand, many gyroscopic systems (e.g., rotating shafts not subjected to axial load) do not exhibit flutter instability at all, the frequencies ω^2 remaining real for the entire range of the parameter (see Section 5.6).

5.4 Gyroscopic conservative systems with multiple paramcters

Considering again the system characterized by (5.18), suppose in addition to ξ^2 there are several more parameters (such as conservative loads) in the system associated with symmetric matrices, and the explicit form of (5.18) is given by

$$(-\omega^2 M + i\omega\xi F + U - \xi^2 E - \eta_k E^k)u = 0 \tag{5.47}$$

where the matrices E^k are symmetric and $k = 1, 2, \ldots, m$.

The characteristic equation

$$\Delta \equiv |-\omega^2 M + i\omega\xi F + U - \xi^2 E - \eta_k F^k| = 0 \tag{5.48}$$

describes $(m+1)$ dimensional characteristic surfaces in the $(m+2)$ dimensional $\omega^2 - \xi^2 - \eta_k$ space. It is, of course, noted that due to skew-symmetry of F and the symmetry of the remaining matrices, the expansion of the determinant (5.48) is an implicit function of ω^2, ξ^2 and η_k, and (5.48) can be written as

$$\Delta(\omega^2, \xi^2, \eta_k) = 0. \tag{5.49}$$

Defining again *the corresponding nongyroscopic system* as the one described by (5.47) with $F = 0$, it is immediately observed that this system is a multiple-parameter conservative system as discussed in Chapter 3. Therefore, according to Theorem 3.1, the fundamental characteristic surface of the system $F = 0$ cannot have convexity toward the origin. It also follows that the fundamental characteristic surface (5.48) in the subspace $\omega^2 - \eta_k$ (i.e., intersection with $\xi^2 = 0$) has the same property

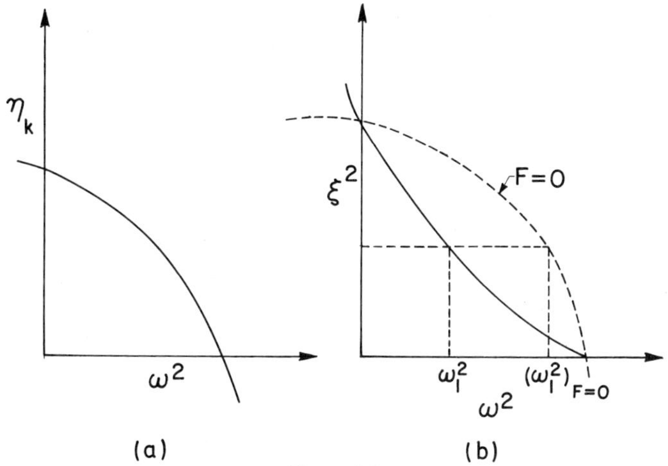

Figure 5.2

(Fig. 5.2(a)) and it coincides with the surface of the $F=0$ system. Hence the following theorem [60] can be stated:

THEOREM 5.2. *The fundamental characteristic surface of a gyroscopic conservative system in the $\omega^2-\eta_k$ subspace cannot have convexity toward the origin.*

On the other hand, in the $\omega^2-\xi^2$ plane (i.e., for all $\eta_k=0$) the two fundamental curves are distinct and obey Theorem 5.1 (Fig. 5.2(b)).

Interest now focusses on the stability properties of the system (5.47) as reflected in the parameter space $\xi^2-\eta_k$. When all the parameters are zero, one has real positive frequencies ω_k^2 and the system is stable. As the values of ξ^2 and η_k are varied, instability may occur by divergence or flutter. Initial instability, however, will always be of the divergence type characterized by $\omega_j^2=0$ for some j, as stated before and implied by (5.8). The locus of all divergence points in the $\xi^2-\eta_k$ space defines the *divergence boundary*, and the following theorem [60] can be formulated:

THEOREM 5.3. *The intial divergence boundary cannot have convexity toward the primary region of stability.*

This theorem is again analogous to and is based on Theorem 3.2.

Depending on the values of η_k, gyroscopic forces may or may not stabilize the system which appears to be unstable in the absence of such forces. Figure 5.3 shows the stabilizing effect of gyroscopic forces. As Fig. 5.3(b) illustrates, it may happen that the characteristic curve in the $\omega^2-\xi^2$

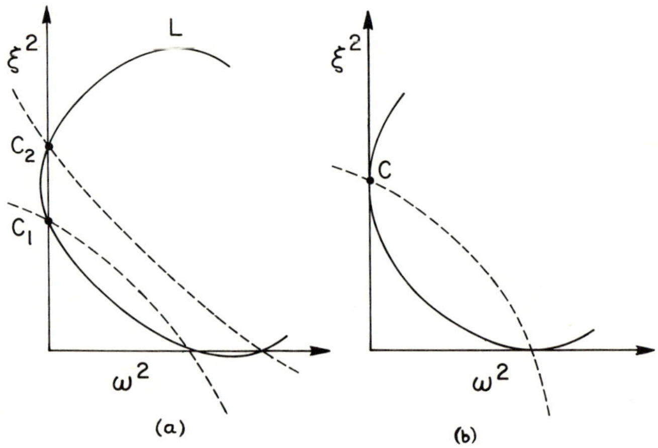

Figure 5.3

plane is tangential to the ξ^2 axis at the critical point C, the system being stable for both $\xi^2 > \xi_c$ and $\xi^2 < \xi_c$ at least in the vicinity of C. A part of the divergence boundary may, then, lie between two stable regions while the remaining part associated with η_k demarcates stable and unstable regions. This distinction arises close to the η_k axes where the characteristic surface tends to change curvature and approach that of the corresponding nongyroscopic system.

For some combinations of the parameters, the characteristic surface may have extrema where two frequencies ω_j^2 coalesce. At such a point the system loses its stability by flutter, and the locus of all such points in the $\xi^2 - \eta_k$ space defines the *flutter boundary*. Analytically, the flutter boundary is also defined by the simultaneous equations (5.49) and

$$\frac{\partial\Delta(\omega^2, \xi^2, \eta_k)}{\partial\omega^2} = 0 \tag{5.50}$$

as was shown in Section 2.5.

It will be seen later that the flutter boundary may have convexity properties in contrast with divergence boundary.

5.5 Standard forms of eigenvalue problems

As emphasized in the preceding sections, the characteristic equation of a gyroscopic conservative system is a function of ω^2 (and ξ^2) like nongyroscopic conservative systems. One might, therefore, enquire whether the original mathematical formulation describing the dynamic behaviour and involving the skew-symmetric matrix G can be transformed into a standard eigenvalue problem in which ω^2 appears as an eigenvalue. Such a transformation has conceptual as well as computational advantages. For example, it enables the analyst to use a large variety of existing computational algorithms in the solution of the eigenvalue problem.

Two types of standard forms [61], concerning the gyroscopic systems (5.18) and (5.21), will be presented here. In the former case, (5.18) will take the standard form in which ω does not appear and in the latter case, (5.21) will be transformed into a *double eigenvalue problem* in ω^2 and ξ^2.

Let (5.18) be represented by

$$(A-\omega I)z=0 \tag{5.51}$$

where

$$A=\begin{bmatrix} iM^{-1}G & M^{-1}K \\ I & 0 \end{bmatrix} \quad \text{and} \quad z=\begin{bmatrix} \omega u \\ u \end{bmatrix}. \tag{5.52}$$

This problem can in turn be transformed into one involving a Hermitian matrix pencil. Thus, on pre-multiplying (5.51) by

$$S_1=\begin{bmatrix} M & 0 \\ 0 & K \end{bmatrix} \tag{5.53}$$

one obtains

$$(S_2-\omega S_1)z=0 \tag{5.54}$$

where

$$S_2=S_1A=\begin{bmatrix} iG & K \\ K & 0 \end{bmatrix} \tag{5.55}$$

and both S_1 and S_2 are Hermitian.

Suppose first that the system is operating in the first stability region so that K is positive definite. It follows immediately that A is symmetrizable and all the eigenvalues ω are real in this region as expected. The eigenvalues ω are the stationary values of the Rayleigh quotient

$$R(x,\bar{x})=\frac{\langle \bar{x}, S_2x\rangle}{\langle \bar{x}, S_1x\rangle} \tag{5.56}$$

However, as remarked earlier, the characteristic polynomial of the gyroscopic system (5.18) is a function of ω^2, and (5.51) does not therefore seem to be an appropriate form. Furthermore, (5.51) does not lend itself to a meaningful comparison with the corresponding nongyroscopic system. To achieve the desired form, consider [61, 66] the matrix pencil

$S_2+\omega S_1$ and pre-multiply (5.54) first by S_1^{-1} and then by $S_2+\omega S_1$ to get

$$(S_2+\omega S_1)S_1^{-1}(S_2-\omega S_1)z=0 \tag{5.57}$$

which reduces to

$$(S_3-\omega^2 S_1)z=0 \tag{5.58}$$

where

$$S_3=S_2S_1^{-1}S_2. \tag{5.59}$$

The eigenvalues of the original problem appear twice as the eigenvalues of (5.58). Using (5.53), (5.55) and (5.59) one obtains

$$S_3=\begin{bmatrix} K-GM^{-1}G & iGM^{-1}K \\ iKM^{-1}G & KM^{-1}K \end{bmatrix}. \tag{5.60}$$

Here both S_1 and S_3 are Hermitian while S_1 is also positive definite. The Rayleigh Quotient

$$R(x,\bar{x})=\frac{\langle \bar{x}, S_3x\rangle}{\langle \bar{x}, S_1x\rangle} \tag{5.61}$$

satisfies

$$\omega_1^2\le R(x,\bar{x})\le \omega_{2n}^2 \tag{5.62}$$

where the roots are assumed to be in ascending order. If the eigenvector of the corresponding nongyroscopic system associated with the smallest frequency $(\omega_1^2)_{G=0}$ is denoted by u^1 (which is now real), then (5.62) with $x=(0,u^{1\prime})'$ yields

$$\omega_1^2\le(\omega_1^2)_{G=0}$$

upon noting that $M^{-1}Ku^1=(\omega_1^2)_{G=0}u^1$. This is another proof of Theorem 5.1.

It is noted that (5.58) can be converted into finding the eigenvalues of the Hermitian matrix

$$C=S_1^{-\frac{1}{2}}S_3S_1^{-\frac{1}{2}}. \tag{5.63}$$

So far it was assumed that the matrix K is positive definite, and some operations leading to (5.58) are based on this condition. Nevertheless, the matrices S_1, S_2 and S_3 are by no means unique and when the matrix K ceases to be positive definite it is possible to change the structure of the matrix S_1 which in turn will result in a change in the structures of S_2 and S_3. Suppose S_1 is selected as

$$S_1=\begin{bmatrix} M & 0 \\ 0 & M \end{bmatrix}; \tag{5.64}$$

it then follows from (5.52) and (5.55) that

$$S_2 = \begin{bmatrix} iG & K \\ M & 0 \end{bmatrix}, \tag{5.65}$$

while (5.59) yields

$$S_3 = \begin{bmatrix} K - GM^{-1}G & iGM^{-1}K \\ iG & K \end{bmatrix}. \tag{5.66}$$

The eigenvalues ω^2 cannot of course be expected to be necessarily real now, and S_3 is generally asymmetric if S_1 is selected as a positive definite matrix.

In order to transform (5.21) into a double eigenvalue problem, introduce $K = U - \xi^2 E$ and $G = \xi F$ into (5.66) which then takes the form

$$S_3 = \begin{bmatrix} U - \xi^2 E - \xi^2 FM^{-1}F & i\xi FM^{-1}(U - \xi^2 E) \\ i\xi F & U - \xi^2 E \end{bmatrix}. \tag{5.67}$$

The eigenvalue problem

$$(S_3 - \omega^2 S_1)z = 0 \quad \text{where} \quad z = \begin{bmatrix} z_1 \\ z_2 \end{bmatrix} \equiv \begin{bmatrix} \omega u \\ u \end{bmatrix}$$

can now be written as

$$(-\omega^2 S_1 + S_4 - \xi^2 S_5)p = 0 \tag{5.68}$$

where

$$S_4 = \begin{bmatrix} U & iFM^{-1}U \\ 0 & U \end{bmatrix}, \quad S_5 = \begin{bmatrix} E + FM^{-1}F & iFM^{-1}E \\ -iF & E \end{bmatrix} \quad \text{and}$$

$$p = \begin{bmatrix} z_1 \\ \xi z_2 \end{bmatrix}. \tag{5.69}$$

Another interesting form can be generated [61] by pre-multiplying (5.68) by

$$T = \begin{bmatrix} I & -iFM^{-1} \\ 0 & I \end{bmatrix} \tag{5.70}$$

in which case one obtains

$$(-\omega^2 T_1 + T_2 - \xi^2 T_3)p = 0 \tag{5.71}$$

with

$$T_1 = \begin{bmatrix} M & -iF \\ 0 & M \end{bmatrix}, \quad T_2 = \begin{bmatrix} U & 0 \\ 0 & U \end{bmatrix} \quad \text{and} \quad T_3 = \begin{bmatrix} E & 0 \\ -iF & E \end{bmatrix}, \tag{5.72}$$

(5.68) and (5.71) describe a *double eigenvalue problem* [61] as in the case of conservative systems except that the matrices involved are now general matrices. For $\xi^2=0$, (5.71), for example, yields the natural frequencies, the spectrum of these frequencies being repeated twice. As ξ^2 is increased from zero, instability may occur by divergence when one of the roots ω^2 becomes zero or by flutter when two distinct roots coalesce. The latter phenomenon is characterized by

$$\frac{\partial \xi^2}{\partial \omega^2}=0. \tag{5.73}$$

Associating now the system (5.71) with its adjoint

$$r'(-\omega^2 T_1+T_2-\xi^2 T_3)=0, \tag{5.74}$$

one can derive a flutter condition. Thus, differentiating (5.74) with respect to ω^2 and post-multiplying by p yields

$$\left\langle r, \left(-T_1-\frac{\partial \xi^2}{\partial \omega^2}\,T_3\right)p\right\rangle=0 \tag{5.75}$$

which upon making use of (5.73) results in the simple flutter condition

$$\langle r, T_1 p\rangle=0. \tag{5.76}$$

It can be shown that (5.76) is equivalent to the flutter condition

$$2\omega\langle v, Mu\rangle-i\xi\langle v, Fu\rangle=0 \tag{5.77}$$

which was given in (5.43) for $M\equiv I$. To do this, it suffices to observe that the left eigenvector in (5.74) is in the form of

$$r=\begin{bmatrix}\xi v\\ \omega v\end{bmatrix}; \tag{5.78}$$

then

$$\langle r, T_1 p\rangle=2\omega\langle v, Mu\rangle-i\xi\langle v, Fu\rangle=0 \tag{5.79}$$

follows immediately.

A direct variational technique for the approximate determination of critical flutter and divergence loads can then be adopted [67], the method being direct in the sense that it dispenses with the frequencies of vibrations. Thus, following a continuum formulation of a given gyroscopic system, both divergence loads (ξ_d^2) and flutter loads (ξ_f^2) can be obtained on the basis of the quotient

$$\frac{1}{\xi^2}=\frac{\langle r, T_3 p\rangle}{\langle r, T_2 p\rangle} \tag{5.80}$$

which is stationary with respect to kinematically admissible variations in p and r, variations being subject to the *constraint* (5.76) in the latter case (ξ_f^2). Details of this method, including illustrative examples, will be presented in Chapter 6.

As described in Section 5.2, the general gyroscopic system (5.71) can represent a pipe with internal flow as well as a rotating shaft, and in the latter case the matrices E in T_3 are replaced by M resulting in

$$T_1 = \bar{T}_3'. \tag{5.81}$$

For rotating systems, one then has in the region of real ω^2

$$\bar{p}'(-\omega^2 T_3 + T_2 - \xi^2 T_1) = 0 \tag{5.82}$$

which is obtained by taking the conjugate transpose of (5.71). It is now observed that $\bar{p}$ is the left eigenvector of the problem in which ω^2 and ξ^2 exchange roles. Pre-multiplying (5.71) by $\bar{p}$ and post-multiplying (5.82) by p and subtracting yields

$$(-\omega^2 + \xi^2)(\langle \bar{p}, T_3 p\rangle - \langle \bar{p}, T_1 p\rangle) = 0$$

which indicates that

$$\langle \bar{p}, T_3 p\rangle = \langle \bar{p}, T_1 p\rangle \quad \text{for} \quad \omega^2 \neq \xi^2.$$

It also follows from (5.81) that the above quantities are real, and (5.82) therefore yields

$$\omega^2 + \xi^2 = \frac{\langle \bar{p}, T_2 p\rangle}{\langle \bar{p}, T_1 p\rangle}. \tag{5.83}$$

5.6 Rotating systems

The formulation in the preceding sections of this chapter is general and not limited to rotating systems. Considering the famous "wheel" problem, however, it is safe to assume that gyroscopic systems actually originated with rotating systems. In this special case, certain restrictions can be introduced on the general gyroscopic system as remarked in Section 5.2 and further more specific results may be expected.

First consider a particle of mass m rotating with a constant angular velocity ξ (e.g., a simple model of a massless elastic shaft with a disk). The equations of motion of this system can easily be derived either using stationary coordinates (x, y) or introducing a coordinate system (p, q) which rotates with the same angular velocity ξ (Fig. 5.4). The position vector in stationary and rotating coordinates is given by

$$r = xe_x + ye_y \tag{5.84}$$

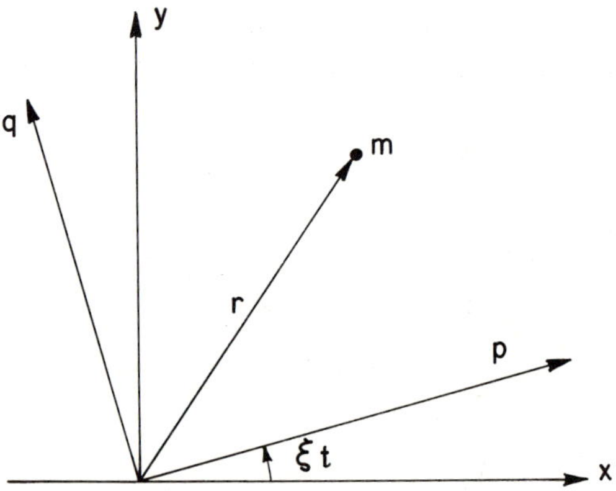

Figure 5.4

and

$$r = pe_p + qe_q \tag{5.85}$$

respectively, where e_x, e_p, etc. are the corresponding unit vectors.

A rotating system can be identified as a gyroscopic system only when treated in a rotating reference frame, and this approach will be followed here. The stationary coordinates, however, will be useful later for introducing external damping.

The velocity and acceleration of the mass m are given by

$$\dot{r} = (\dot{p} - \xi q)e_p + (\dot{q} + \xi p)e_q \tag{5.86}$$

and

$$\ddot{r} = (\ddot{p} - 2\xi\dot{q} - \xi^2 p)e_p + (\ddot{q} + 2\xi\dot{p} - \xi^2 q)e_q \tag{5.87}$$

respectively where $-2\xi\dot{q}$, $2\xi\dot{p}$ are Coriolis components of the acceleration.

If the forces acting on the particle are assumed to consist of elastic restoring forces (the effect of damping will be studied later), one has

$$\left.\begin{aligned} &F = -c_1 x e_x - c_2 y e_y \\ \text{or}\quad &\\ &F = -c_1 p e_p - c_2 q e_q \end{aligned}\right\} \tag{5.88}$$

and the application of Newton's Law yields the linear equations of motion in stationary coordinates as

$$\begin{aligned} &m\ddot{x} + c_1 x = 0 \\ &m\ddot{y} + c_2 x = 0 \end{aligned} \tag{5.89}$$

and in rotating coordinates as

$$\begin{aligned} m\ddot{p} - 2\xi m\dot{q} + (c_1 - \xi^2 m)p = 0 \\ m\ddot{q} + 2\xi m\dot{p} + (c_2 - \xi^2 m)q = 0. \end{aligned} \tag{5.90}$$

Considering n particles, and an additional conservative parameter η (e.g., an axial conservative load acting on the shaft), one can readily generalize (5.90) to obtain the linearized equations of motion of a rotating system as

$$\left.\begin{aligned} M\ddot{p} - 2\xi M\dot{q} + (U - \xi^2 M - \eta E)p = 0 \\ M\ddot{q} + 2\xi M\dot{p} + (V - \xi^2 M - \eta E)q = 0 \end{aligned}\right\} \tag{5.91}$$

where $M = M' > 0$, $U = U' > 0$, $V = V' > 0$ and $E = E'$. If the elastic restoring forces (or flexural rigidities) in two principal directions are the same, as in the case of a circular shaft, then $U = V$. The column vectors p and q represent deflections in two orthogonal directions in the rotating frame. One can introduce $p = p_0 e^{i\omega t}$ and $q = q_0 e^{i\omega t}$ to obtain

$$\left\{ -\omega^2 \begin{bmatrix} M & 0 \\ 0 & M \end{bmatrix} + 2i\omega\xi \begin{bmatrix} 0 & -M \\ M & 0 \end{bmatrix} + \begin{bmatrix} U & 0 \\ 0 & V \end{bmatrix} - \xi^2 \begin{bmatrix} M & 0 \\ 0 & M \end{bmatrix} - \eta \begin{bmatrix} E & 0 \\ 0 & E \end{bmatrix} \right\} \begin{bmatrix} p_0 \\ q_0 \end{bmatrix} = 0 \tag{5.92}$$

which is a special case of (5.47) with the characteristic equation

$$\begin{vmatrix} U - \eta E - (\xi^2 + \omega^2)M & -2i\omega\xi M \\ +2i\omega\xi M & V - \eta E - (\xi^2 + \omega^2)M \end{vmatrix} = 0 \tag{5.93}$$

The uniform system with $U = V$ and the asymmetric system with $U \neq V$ will be studied separately.

(i) *Uniform system: $U = V$*

In this case, introducing the complex displacement $r = p + iq$, (5.91) can simply be written as

$$M\ddot{r} + 2i\xi M\dot{r} + (U - \xi^2 M - \eta E)r = 0. \tag{5.94}$$

Assuming $r = ue^{i\omega t}$ yields

$$(-\omega^2 M - 2\xi\omega M + U - \xi^2 M - \eta E)u = 0 \tag{5.95}$$

and pre-multiplying by $\bar{u}'$ results in

$$\omega = \frac{+\xi m \pm \sqrt{m(a-\eta e)}}{-m} \tag{5.96}$$

where $m = \langle \bar{u}, Mu \rangle$, $a = \langle \bar{u}, Uu \rangle$ and $e = \langle \bar{u}, Eu \rangle$.

Clearly, if $\eta = 0$, ω will always be real and flutter instability is then ruled out. In other words, an elastic shaft (for example) cannot lose stability by flutter if it is not subjected to external forces. Furthermore, divergence can only occur at the critical points $\xi^2 = a/m$. In the presence of axial force, the flutter boundary, $a - \eta e = 0$, consists of straight lines $\eta = \eta_j$ $(j = 1, 2, \ldots, n)$ parallel to the ξ^2 axis where η_j are the critical divergence points on the η axis and all eigenvectors are taken into account. In other words, observing that $U - \eta E$ is positive definite for sufficiently small η, when $a - \eta e = 0$ one has

$$|U - \eta E| = 0. \text{ (flutter boundary).} \tag{5.97}$$

The flutter frequency is obtained from (5.96) as

$$\omega_f = -\xi. \tag{5.98}$$

The flutter boundary (5.97) can also be derived directly from (5.93). To this end, if one adds i times the last n rows to the first n rows, and then substracts i times the first n columns from the last n columns, respectively, one obtains

$$\begin{vmatrix} U - \eta E - (\xi + \omega)^2 M & i(U - V) \\ 2i\omega\xi M & V - \eta E - (\xi - \omega)^2 M \end{vmatrix} = 0 \tag{5.99}$$

which, for $U = V$, leads to

$$|U - \eta E - (\xi + \omega)^2 M| = 0 \tag{5.100}$$

or

$$|U - \eta E - (\xi - \omega)^2 M| = 0. \tag{5.101}$$

Since the matrices in these equations are symmetric and M is positive definite, the roots $(\xi + \omega)^2$ and $(\xi - \omega)^2$ are real. Considering now, for instance, the eigenvalue problem corresponding to (5.100),

$$[U - \eta E - (\xi + \omega)^2 M]u = 0,$$

one has

$$\langle u, [U - \eta E - (\xi + \omega)^2 M]u \rangle = 0$$

which upon differentiating with respect to ω yields

$$\left\langle u, \left[-\frac{\partial \eta}{\partial \omega} - 2(\xi + \omega)\left(\frac{\partial \xi}{\partial \omega} + 1\right) M \right] u \right\rangle = 0$$

and at a flutter point

$$(\xi + \omega)\langle u, Mu \rangle = 0$$

or

$$\omega + \xi = 0.$$

Using this result, (5.100) yields the flutter boundary.

The divergence boundary is readily obtained as

$$|U - \eta E - \xi^2 M| = 0, \tag{5.102}$$

and according to Theorem 5.3 the initial divergence boundary cannot have convexity toward the origin. In the special case when the matrices U, E and M can simultaneously be reduced to diagonal form (e.g., simply supported rotating shafts), the divergence curves described by (5.102) consist of straight lines.

On the basis of the foregoing discussion, the following results concerning the stability of uniform ($U = V$) rotating systems can be stated:

(a) THEOREM 5.4. *A uniform rotating system cannot lose stability by flutter if the additional external force η is absent.*

(b) THEOREM 5.5. *The flutter boundary of a uniform rotating system consists of straight lines $\eta = \eta_j$ parallel to the ξ^2 axis where η_j are the divergence points on the η axis.*

(c) *The initial divergence boundary of a uniform rotating system cannot have convexity toward the primary region of stability (Theorem 5.3), and if $\eta = 0$ divergence is exhibited only at the critical points. There are no regions of divergence.*

(ii) *Asymmetric systems: $U \neq V$*

Consider now the more general case of nonuniform rotating systems. It follows immediately from (5.93) that the divergence boundary is given by

$$|U - \eta E - \xi^2 M| = 0$$

and (5.103)

$$|V - \eta E - \xi^2 M| = 0.$$

In the majority of practical systems V happens to be simply a multiple of U, and the discussion concerning the divergence boundary of uniform systems applies here as well.

In many applications when V is a multiple of U, the matrices M, U, and E can be reduced to diagonal matrices simultaneously. In this case, (5.93) becomes the product of (2×2) determinants of the form

$$\begin{vmatrix} u-\eta e-(\xi^2+\omega^2)m & -2i\omega\xi m \\ 2i\omega\xi m & v-\eta e-(\xi^2+\omega^2)m \end{vmatrix} = 0 \tag{5.104}$$

where small letters are used to denote the elements of the corresponding matrices. There are n equations like (5.104), and it can easily be verified that for $\eta = 0$ flutter instability is again ruled out since the discriminant of the equation in ω^2 is always positive, i.e.,

$$(u-v)^2+8m\xi^2(u+v)>0.$$

The condition for a double root ω^2 is found to be

$$\xi^2 = \frac{(u-v)^2}{8m(2\eta e-u-v)} \quad (n \text{ equations}) \tag{5.105}$$

which describes the flutter boundary. These curves are branches of hyperbolas with horizontal asymptote

$$\xi^2 = 0 \tag{5.106}$$

and vertical asymptote

$$\eta = \frac{u+v}{2e} \tag{5.107}$$

if $e \neq 0$. They are tangential to the divergence boundary (5.103) at

$$\xi^2 = \frac{u-v}{4m}. \tag{5.108}$$

The following results can then be stated for the system under consideration:

(a) THEOREM 5.6. *The asymmetric rotating system cannot exhibit flutter instability if the additional external force η is absent.*

(b) THEOREM 5.7. *The flutter boundary of the asymmetric rotating system consists of branches of hyperbolas with η axis as a horizontal asymptote and with vertical asymptotes midway between the pairs of critical divergence points on the η axis.*

(c) *The initial divergence boundary of an asymmetric rotating system falls within the scope of Theorem 5.3 as in the case of uniform systems. For* $\eta = 0$, *however, the system now exhibits divergence regions.*

5.7 Example 1: Rotating model

Consider a simplified model of a disk (with mass m) mounted on a non-circular weightless shaft which is also subjected to a constant axial compression force. Formulating this problem according to Section 5.6, and arranging the terms appropriately, one obtains the gyroscopic system (5.92) in the form

$$\left\{-\omega^2\begin{bmatrix}1 & 0\\ 0 & 1\end{bmatrix}+2i\omega\xi\begin{bmatrix}0 & -1\\ 1 & 0\end{bmatrix}+\begin{bmatrix}c_1-\xi^2-\eta & 0\\ 0 & c_2-\xi^2-\eta\end{bmatrix}\right\}\begin{bmatrix}p_0\\ q_0\end{bmatrix}=0 \tag{5.109}$$

where c_1 and c_2 represent the elastic rigidities in two principal directions and η is the parameter representing axial compression. In deriving (5.109), the shaft is assumed to be rigid with two torsional springs (Fig. 5.5), and the original equations of motion are divided by ml^2 so that $\eta = P/ml$. The rotary inertia of the disk as well as the weight of the shaft is neglected.

Consider first the case of $\eta = 0$; the characteristic equation is then given by

$$(\omega^2)^2-(c_1+c_2+2\xi^2)\omega^2+(c_1-\xi^2)(c_2-\xi^2)=0. \tag{5.110}$$

Ziegler [13] has used Vieta's relations to determine the unstable region and to explain the stabilizing effect of gyroscopic forces. It seems more illustrative to plot the characteristic curve in the $\omega^2-\xi^2$ plane as shown in Figure 5.6(i) for the case $c_1 = 2$, $c_2 = 6$. The curve is a parabola which is symmetric with respect to the bisector of the first quadrant. Stability is

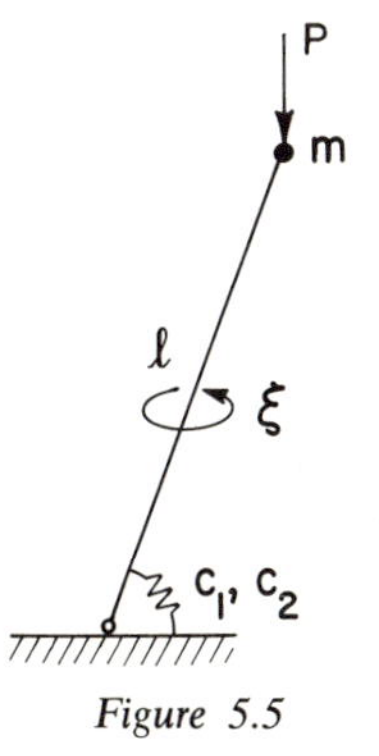

Figure 5.5

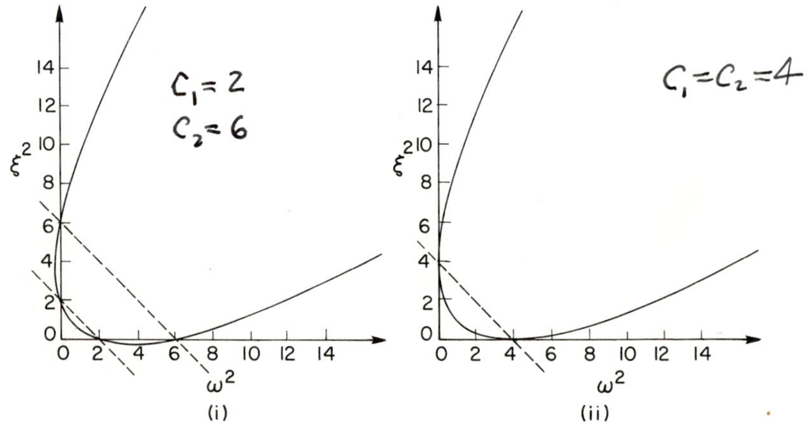

Figure 5.6

lost by divergence at $\xi^2 = c_1$ and the system remains unstable until $\xi^2 = c_2$. With further increase in angular velocity, stability is regained since the curve then lies in the stable region $\omega^2 > 0$. The characteristic curves for the corresponding non-gyroscopic system are depicted by dashed lines in the figures, and in this case they are straight lines.

If the elastic rigidities of the shaft are the same in both principal directions, the characteristic curve for the rotating disk is tangential to the ξ^2 and ω^2 axes at the value c_1. The case $c_1 = c_2 = 4$ is shown in Figure 5.6 (ii). The system is only unstable at $\xi^2 = c_1$; otherwise the characteristic curve lies in the region $\omega^2 > 0$. It is interesting to note that in this special case the stationary shaft ($\xi^2 = 0$), although in space and having two essential kinetic degrees of freedom, behaves like a single degree of freedom system. As soon as rotation starts, however, the effect of gyroscopic Coriolis forces is to expose the actual system by producing two distinct frequencies ω^2 of free transverse vibrations. It is further noted that the fundamental characteristic curve of the corresponding non-gyroscopic system is an upper bound for that of the gyroscopic system as Theorem 5.1 states. According to Theorems 5.4 and 5.6 flutter instability is not expected for $\eta = 0$ and it is clear from the parabolas in Figure 5.6 that the characteristic curves actually cannot have a maximum.

Now let $\eta \neq 0$ and consider first the non-uniform shaft. The divergence boundary is readily obtained from the characteristic equation

$$\Delta = (\omega^2)^2 - (c_1 + c_2 + 2\xi^2 - 2\eta)\omega^2 + (c_1 - \xi^2 - \eta)(c_2 - \xi^2 - \eta) = 0 \tag{5.111}$$

as

$$c_1 - \xi^2 - \eta = 0$$
$$c_2 - \xi^2 - \eta = 0 \tag{5.112}$$

which describe two parallel lines (Fig. 5.7).

For the flutter boundary one has the condition

$$\frac{\partial \Delta}{\partial \omega^2} = 2\omega^2 - (c_1 + c_2 + 2\xi^2 - 2\eta) = 0$$

which together with (5.111) yields the flutter boundary

$$1 + 4\xi^2(5 - \eta) = 0 \tag{5.113}$$

for $c_1 = 4$ and $c_2 = 6$. (5.113) describes a hyperbola (Fig. 5.7) with asymptotes $\xi^2 = 0$ and $\eta = (4+6)/2 = 5$. The flutter boundary is tangent to the divergence boundary at $\xi^2 = (6-4)/4 = 0{\cdot}5$ and $\eta = 5{\cdot}5$. The regions of stability, divergence and flutter are indicated by *S*, *D* and *F* respectively in Figure 5.7.

Considering an arbitrary ray τ in the $\xi^2 - \eta$ parameter space defined by

$$\eta = l_1\tau, \qquad \eta = l_2\tau,$$

the $\tau - \omega^2$ relationship can readily be determined from (5.111). As an example, this has been done for $l_1 = \cos 60°$ and $l_2 = \cos 30°$, and the

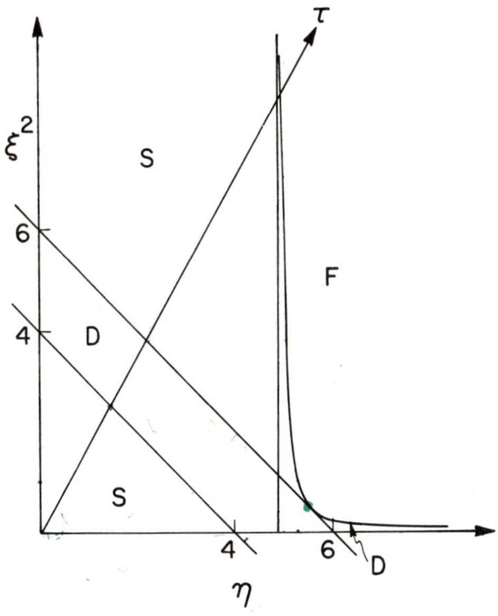

Figure 5.7

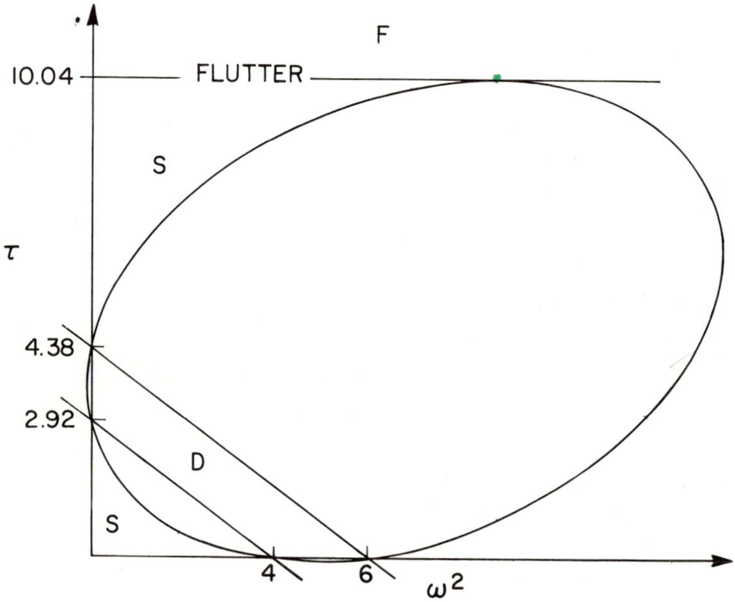

Figure 5.8

result is shown in Figure 5.8. While for $\eta = 0$ the characteristic curve is a parabola, now it is an ellipse which demonstrates clearly how flutter can occur. The ray τ, emerging from the origin of the parameter space, intersects the initial divergence boundary first where the system loses stability by divergence. Upon intersecting the second divergence boundary, however, the system regains stability, and as the ray proceeds in the stable region the flutter boundary is intersected at a point where the characteristic curve has a maximum. Beyond this point, the frequencies ω^2 are obviously complex.

This example can also be used to illustrate some aspects of the theory presented in Section 5.5. The standard form (5.71), for example, can immediately be constructed with the matrices

$$T_1 = \begin{bmatrix} 1 & 0 & 0 & +2i \\ 0 & 1 & -2i & 0 \\ 0 & 0 & 1 & 0 \\ 0 & 0 & 0 & 1 \end{bmatrix}, \quad T_2 = \begin{bmatrix} c_1-\eta & 0 & 0 & 0 \\ 0 & c_2-\eta & 0 & 0 \\ 0 & 0 & c_1-\eta & 0 \\ 0 & 0 & 0 & c_2-\eta \end{bmatrix}$$

and

$$T_3 = \begin{bmatrix} 1 & 0 & 0 & 0 \\ 0 & 1 & 0 & 0 \\ 0 & 2i & 1 & 0 \\ -2i & 0 & 0 & 1 \end{bmatrix}.$$

Consider now an arbitrary point, $\eta = 5{\cdot}25$ and $\xi^2 = 1$ for instance, on the flutter boundary (5.113); the corresponding frequency, which is a repeated root, can readily be determined as $\omega^2 = 0{\cdot}75$. The right and left eigenvectors, p and r, of the double eigenvalue problem (5.71) corresponding to $\omega^2 = 0{\cdot}75$ and $\xi^2 = 1$ (with $\eta = 5{\cdot}25$) now follow from (5.71) and (5.74) respectively. These vectors are of the forms

$$p' = (1 - i,\ 3 + 3i,\ 2 - 2i,\ 2 + 2i),$$

$$r' = (2 + 2i,\ 2 - 2i,\ 1 + i,\ 3 - 3i).$$

The flutter condition (5.76) is now readily verified since

$$\langle r,\ T_1 p\rangle = 0.$$

This condition holds for all points on the flutter boundary.

On the other hand, for the same point on the flutter boundary, the use of p and its complex conjugate $\bar{p}$ yields

$$\frac{\langle \bar{p}, T_2 p\rangle}{\langle \bar{p}, T_1 p\rangle} = \frac{7}{4} = 1{\cdot}75$$

which verifies (5.83) since $\omega^2 + \xi^2 = 0{\cdot}75 + 1{\cdot}00 = 1{\cdot}75$.

Finally, a system with four degrees of freedom is considered. If the uniform model (with $\eta = 0$) is formulated as a four degree of freedom system with two disks, the matrices in the general equation (5.21) will be of the form

$$U = \begin{bmatrix} 2 & 1 & 0 & 0 \\ 1 & 4 & 0 & 0 \\ 0 & 0 & 2 & 1 \\ 0 & 0 & 1 & 4 \end{bmatrix}, \quad G = \begin{bmatrix} 0 & 0 & -2 & 0 \\ 0 & 0 & 0 & -2 \\ 2 & 0 & 0 & 0 \\ 0 & 2 & 0 & 0 \end{bmatrix}, \qquad M = I, \qquad E = I;$$

so this system also falls into category (a) of Section 5.2 as the two-degree-of-freedom system. The characteristic curves are shown in Figure 5.9(i) and represent an extension of the type depicted in Figure 5.6(i). There are two points of divergence on the load axis, at $\xi^2 = 3 \pm \sqrt{2}$; other values of ξ^2 yield stability. The matrices for the system shown in Figure 5.9(ii)

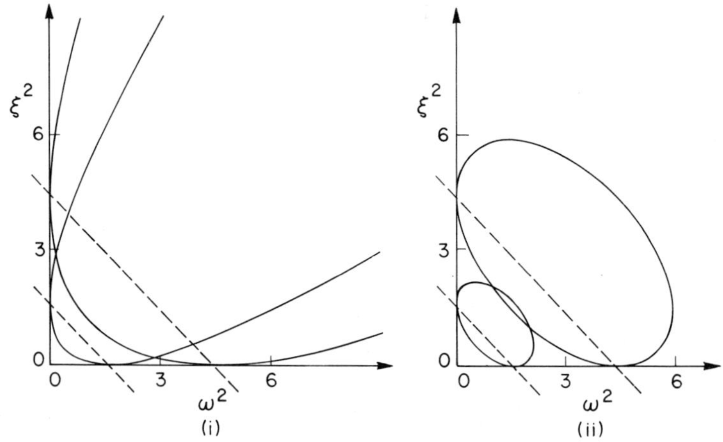

Figure 5.9

are the same as before except that G is half of the one given there. This example is not derived from a physical problem, but the characteristic curves illustrate the flutter phenomenon. The sequence on the load axis is stability, divergence (at a point), stability, flutter, etc.

5.8 Example 2: Rotating flexible shafts

Consider an elastic shaft rotating with a constant angular velocity $\bar{\xi}$ and loaded with a constant axial force $\bar{\eta}$. Let x denote the axial coordinate and $y_j(x, t)$ the transverse displacements in the principal directions ($j = 1, 2$), $s_j = EI_j$ the flexural rigidities in the principal directions, l the length, and μ the mass per unit length of the shaft. A disk of mass $\bar{m}$ may be located at $x = l_0$ where $0 < l_0 < l$, and its displacements are $w_j(t) = y_j(l_0, t)$. Let Δ represent changes of quantities at the disk, e.g.,

$$\Delta w_j(t) = y_j(l_0+, t) - y_j(l_0-, t) \tag{5.114}$$

and it is assumed that

$$\Delta\left(\frac{\partial w_j}{\partial x}\right) = 0. \tag{5.115}$$

Neglecting the effects of rotary inertia and gyroscopic moments caused by the disk, the equations of motion of the shaft are in the form

$$\left.\begin{aligned} \mu\ddot{y}_1 - 2\bar{\xi}\mu\dot{y}_2 + s_1 y_1'''' + \bar{\eta}y_1'' - \bar{\xi}^2\mu y_1 = 0 \\ \mu\ddot{y}_2 + 2\bar{\xi}\mu\dot{y}_1 + s_2 y_2'''' + \bar{\eta}y_2'' - \bar{\xi}^2\mu y_1 = 0 \end{aligned}\right\} \tag{5.116}$$

and the transition conditions at the disk are given by

$$\left.\begin{aligned}&\bar m\ddot w_1-2\bar\xi\bar m\dot w_2+s_1\Delta w_1'''-\bar\xi^2\bar m w_1=0\\&\bar m\ddot w_2+2\bar\xi\bar m\dot w_1+s_2\Delta w_2'''-\bar\xi^2 m w_2=0\end{aligned}\right\}\tag{5.117}$$

$$\Delta w_k''=0,\qquad k=1,2.$$

It will be assumed that the ends of the shaft are either simply supported or clamped so that the boundary conditions at each end are either

$$y_j=0,\qquad y_j''=0\quad\text{or}\quad y_j=0,\qquad y_j'=0;\qquad j=1,2.\tag{5.118}$$

Differential equations (5.116) are within the context of elementary beam theory, and can be derived in various ways. Following the procedure of Section 4.7, for instance, one can write the bending moment (Fig. 5.10) of the forces acting on an element $\mu\,\mathrm{d}\varphi$ at x. In Fig. 5.10 the shaft is shown as simply supported, but the analysis remains essentially the same for the clamped conditions. The forces acting on the element follow from (5.87). Thus, in y_1 direction one has

$$s_1y_1''(x,t)=-R(l-x)-\bar\eta y_1(x,t)-\mu\int_x^l(\varphi-x)\ddot y_1(\varphi,t)\,\mathrm{d}\varphi$$
$$+\mu\bar\xi^2\int_x^l(\varphi-x)y_1(\varphi,t)\,\mathrm{d}\varphi+2\mu\bar\xi\int_x^l(\varphi-x)\dot y_2(\varphi,t)\,\mathrm{d}\varphi.$$

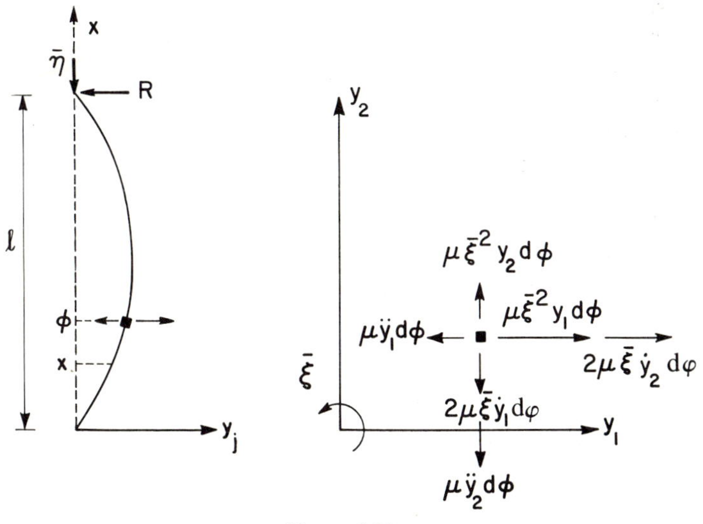

Figure 5.10

Differentiating this equation with respect to x yields

$$s_1 y_1'''(x, t) = R - \bar{n} y_1' \mu \int_x^l \ddot{y}_1(\varphi, t)\, \mathrm{d}\varphi - \mu \bar{\xi}^2 \int_x^l y_1(\varphi, t)\, \mathrm{d}\varphi - 2\mu\bar{\xi} \int_x^l \dot{y}_2(\varphi, t)\, \mathrm{d}\varphi.$$

A second differentiation results in

$$s_1 y_1''''(x, t) = -\bar{\eta} y_1'' - \mu \ddot{y}_1(x, t) + \mu \bar{\xi}^2 y_1(x, t) + 2\mu\bar{\xi}\dot{y}_2(x, t)$$

which is the first of the equations in (5.116). The second equation in (5.116) can similarly be obtained by considering the forces in the y_2 direction.

Assuming motion of the form

$$y_1(x, t) = z_1(x)\mathrm{e}^{i\bar{\omega}t}, \qquad y_2(x, t) = z_2(x)\mathrm{e}^{i\bar{\omega}t} \tag{5.119}$$

(5.116) becomes

$$\begin{aligned} -\bar{\omega}^2 \mu z_1 - 2i\bar{\omega}\bar{\xi}\mu z_2 + s_1 z_1'''' + \bar{\eta} z_1'' - \bar{\xi}^2 \mu z_1 = 0 \\ -\bar{\omega}^2 \mu z_2 + 2i\bar{\omega}\bar{\xi}\mu z_1 + s_2 z_2'''' + \bar{\eta} z_2'' - \bar{\xi}^2 \mu z_2 = 0. \end{aligned} \tag{5.120}$$

Now consider the approximations

$$z_1(x) = a_k \varphi_k(x), \qquad z_2(x) = b_k \varphi_k(x), \qquad k = 1, 2, \ldots, n \tag{5.121}$$

where the functions $\varphi_k(x)$ satisfy the boundary conditions.

Following Galerkin's procedure, (5.119) and (5.121) are applied to the equations of motion, transition conditions (5.117) and boundary conditions (5.118). Multiplying the resulting equations of motion by φ_j and integrating, one gets

$$\left[s_1 \int_0^l \varphi_j \varphi_k'''' \,\mathrm{d}x + \bar{\eta} \int_0^l \varphi_j \varphi_k'' \,\mathrm{d}x - (\bar{\xi}^2 + \bar{\omega}^2)\mu \int_0^l \varphi_j \varphi_k \,\mathrm{d}x \right] a_k - 2i\bar{\omega}\bar{\xi}\mu \int_0^l \varphi_j \varphi_k \,\mathrm{d}x b_k = 0$$

$$\left[s_2 \int_0^l \varphi_j \varphi_k'''' \,\mathrm{d}x + \bar{\eta} \int_0^l \varphi_j \varphi_k'' \,\mathrm{d}x - (\bar{\xi}^2 + \bar{\omega}^2)\mu \int_0^l \varphi_j \varphi_k \,\mathrm{d}x \right] b_k + 2i\bar{\omega}\bar{\xi}\mu \int_0^l \varphi_j \varphi_k \,\mathrm{d}x a_k = 0.$$

Note that the summation in these equations is over $k = 1, 2, \ldots, n$.

Now integrating by parts and using boundary and transition conditions, one obtains (5.92) with

$$u_{jk} = s_1 \int_0^l \phi_j'' \phi_k'' \, dx, \qquad v_{jk} = s_2 \int_0^l \phi_j'' \phi_k'' \, dx$$

$$m_{jk} = \mu \int_0^l \varphi_j \varphi_k \, dx + \bar{m} \varphi_j(l_0) \varphi_k(l_0)$$

$$e_{jk} = \int_0^l \varphi_j' \varphi_k' \, dx, \qquad p_0 = a_k, \qquad q_0 = b_k$$

where u_{jk}, v_{jk}, etc. are the elements of the matrices U, V, etc., and η, ξ, ω are replaced by $\bar{\eta}$, $\bar{\omega}$, and $\bar{\xi}$ respectively. Define the non-dimensional quantities

$$\begin{aligned} &x \to \frac{x}{l}, \qquad \psi_j(x) = \frac{\varphi_j(lx)}{l} \\ &\xi = \frac{\bar{\xi} l^2}{\pi^2} \sqrt{\frac{\mu}{s_1}}, \qquad \eta = \frac{\bar{\eta} l^2}{\pi^2 s_1}, \qquad \omega = \frac{\bar{\omega} l^2}{\pi^2} \sqrt{\frac{\mu}{s_1}} \\ &x_0 = \frac{l_0}{l}, \qquad m = \frac{\bar{m}}{\mu l}, \qquad \alpha = \frac{s_2}{s_1}. \end{aligned} \tag{5.122}$$

If primes now denote differentiation with respect to the new $x(=x/l)$, one then finds

$$\begin{aligned} u_{jk} &= \frac{1}{\pi^4} \int_0^1 \psi_j'' \psi_k'' \, dx, \\ v_{jk} &= \frac{\alpha}{\pi^4} \int_0^1 \psi_j'' \psi_k'' \, dx = \alpha u_{jk}, \\ m_{jk} &= \int_0^1 \psi_j \psi_k \, dx + m \psi_j(x_0) \psi_k(x_0), \\ e_{jk} &= \frac{1}{\pi^2} \int_0^1 \psi_j' \psi_k' \, dx. \end{aligned} \tag{5.123}$$

More generally, with disks of mass $m_1, m_2, \ldots, m_r$ at $x = x_1, \ldots, x_r$,

$$m_{jk} = \int_0^1 \psi_j \psi_k \, dx + \sum_{\beta=1}^{r} m_\beta \psi_j(x_\beta) \psi_k(x_\beta).$$

The modes of free vibration can now be introduced as the functions φ_j to obtain numerical results. In the case of a simply supported shaft, one has

$$\varphi_k = \sin\frac{k\pi x}{l} \qquad k = 1, 2, \ldots$$

Take the first two modes corresponding to $k = 1, 2$

$$\varphi_1 = \sin\frac{\pi x}{l}, \qquad \varphi_2 = \sin\frac{2\pi x}{l}.$$

Using x for x/l results in

$$\psi_1 = \frac{1}{l}\sin \pi x, \qquad \psi_2 = \frac{1}{l}\sin 2\pi x$$

and hence

$$U = \frac{1}{l^2}\begin{bmatrix} 0{\cdot}5 & 0 \\ 0 & 8 \end{bmatrix}, \qquad V = \frac{\alpha}{l^2}\begin{bmatrix} 0{\cdot}5 & 0 \\ 0 & 8 \end{bmatrix},$$
$$E = \frac{1}{l^2}\begin{bmatrix} 0{\cdot}5 & 0 \\ 0 & 2 \end{bmatrix}, \qquad M = \frac{1}{l^2}\begin{bmatrix} 0{\cdot}5 + m & 0 \\ 0 & 0{\cdot}5 \end{bmatrix}, \tag{5.124}$$

for $x_0 = 0{\cdot}5$.

Since all the matrices are diagonal (in this problem), the two vibration modes are independent of each other, and the divergence and flutter boundaries can be obtained for each mode by using (5.102) and (5.97) in the uniform case ($\alpha = 1$) and (5.103) and (5.105) in the asymmetric case.

Computations were carried out [60] for a shaft having one end clamped [103] and the other end simply supported (B) as well as for a shaft simply supported at both ends (A). Various combinations of $\alpha = 1$ and $\alpha = 0{\cdot}5$, $m = 0$ and $m = 1$ at $x_0 = 0{\cdot}5$ (i.e., a disk in the center of the shaft with mass equal to the shaft mass) are considered. The resulting stability and instability regions are depicted in Figures 5.11–5.18 as follows:

Figure	*Ends*	α	m
5.11	A	1	0
5.12	A	1	1
5.13	A	0·5	0
5.14	A	0·5	1
5.15	B	1	0
5.16	B	1	1
5.17	B	0·5	0
5.18	B	0·5	1

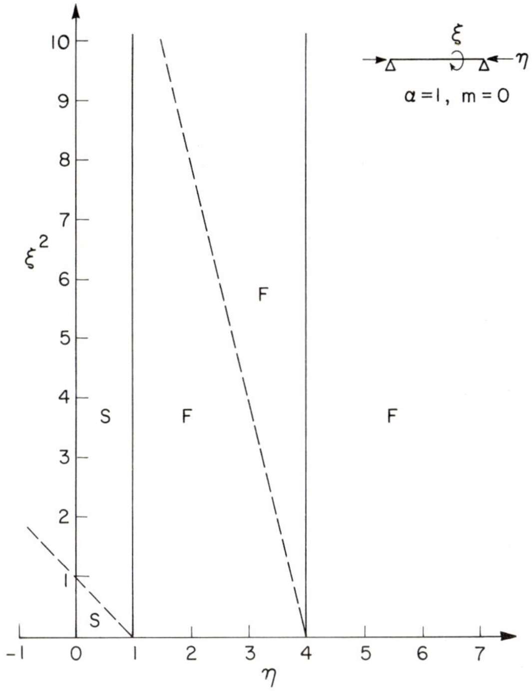

Figure 5.11

S denotes a region of stability, *D* a region in which divergence instability occurs, and *F* designates a region of flutter instability. The divergence boundary is shown with dashed lines and the flutter boundary with solid lines.

It must be remarked that the terminology adopted here in the context of the general theory may not be quite the same as it is commonly used in the literature concerning rotating shafts. In the divergence region, the amplitude of the corresponding mode increases monotonically, and the system may be said to be in "whirling divergence". The flutter instability, on the other hand, appears as self-excited oscillations with increasing amplitude *in the rotating reference frame.*

For Figures 5.11, 5.12, 5.15 and 5.16, $\alpha = 1$ ($U = V$) and hence the branches of the flutter boundary are vertical straight lines. Flutter instability may occur everywhere to the right of the first vertical line (e.g., for $\eta \geq 1$ and $\xi^2 > 0$ in Figures 5.11 and 5.12). The divergence boundary consists of two branches; in Figures 5.11 and 5.12 these branches are straight lines since *M, E, U* and *V* are all diagonal matrices while in Figures 5.15 and 5.16 the initial divergence boundary is concave toward

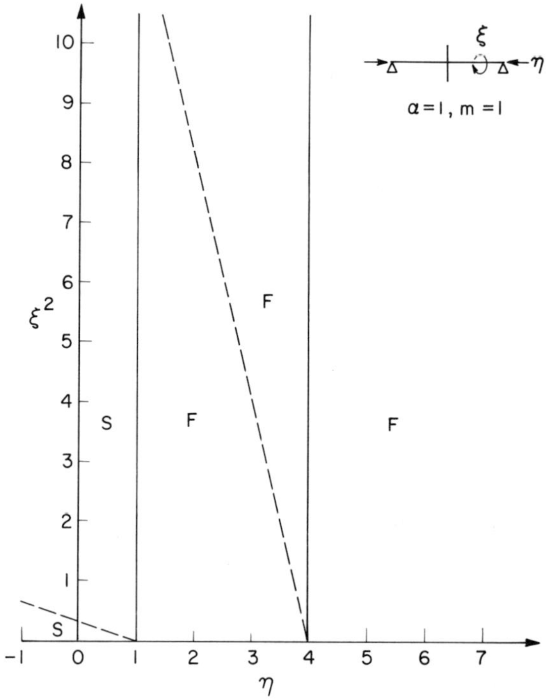

Figure 5.12

the origin (Theorem 5.3). There are no regions of divergence instability in these cases, i.e., divergence only occurs on these branches. The presence of the disk has no effect on the flutter boundary in these cases, but it tends to bring the initial divergence boundary closer to the origin.

In the remaining figures the shaft is asymmetric ($U \neq V$) and the divergence boundary consists of two pairs of branches. A divergence region exists between each of these pairs, and a branch of the flutter boundary is tangential to each of these regions. Divergence also occurs in the regions which are under the flutter branches and to the right of the divergence branches.

The vertical asymptotes of the flutter boundary as $\xi^2 \to 0$ occur midway between pairs of divergence points on the η axis. In Figures 5.13 and 5.14, for example, the asymptotes are $\eta = 0{\cdot}75$ and $\eta = 3$. In these two cases the flutter boundary consists of branches of hyperbolas (since M, E, U and V are diagonal) and no flutter is possible for $\eta < 0{\cdot}75$ including $\eta = 0$ (Theorem 5.6). The leftmost vertical asymptote in Figures 5.17 and 5.18 also provides a bound on flutter; however, this property is

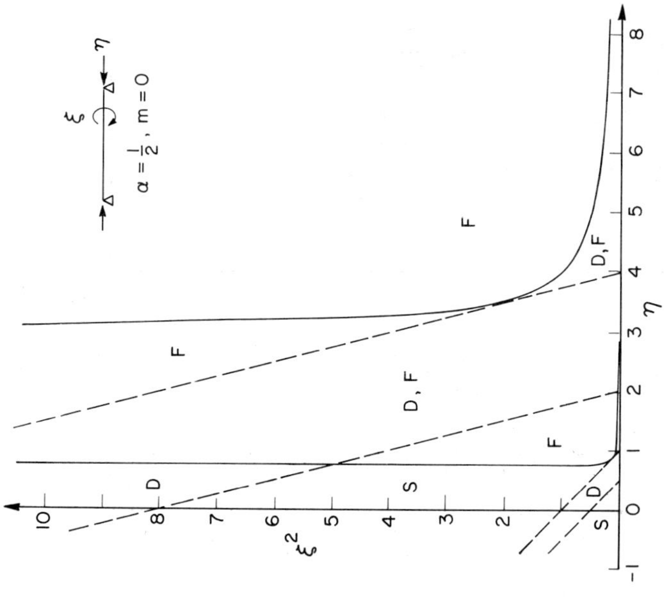

Figure 5.13

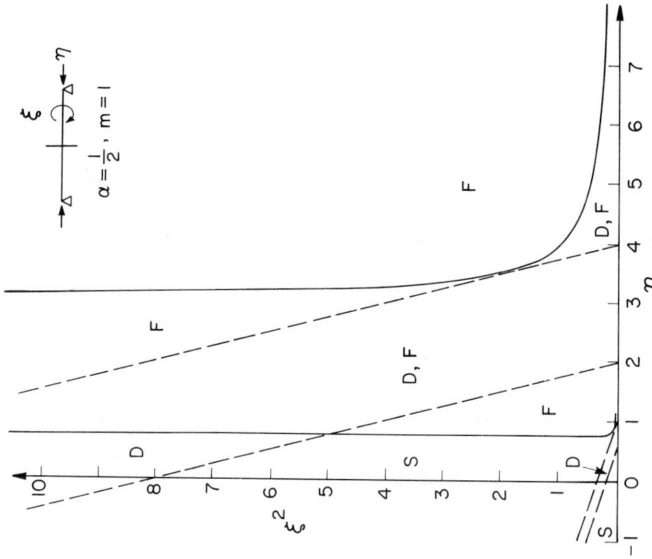

Figure 5.14

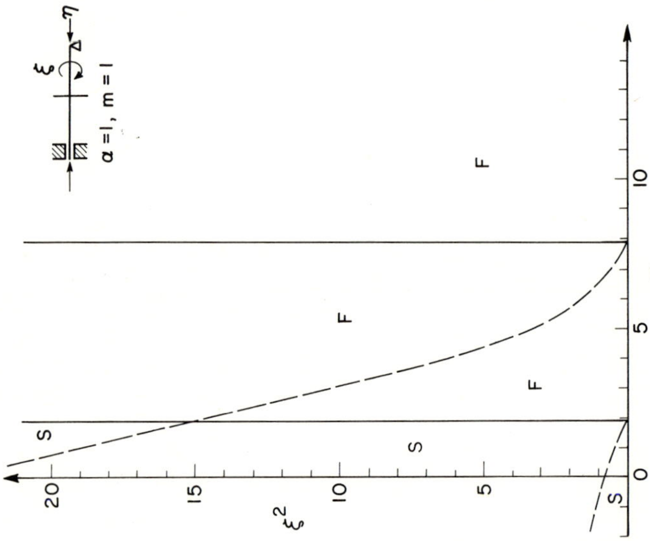

Figure 5.16

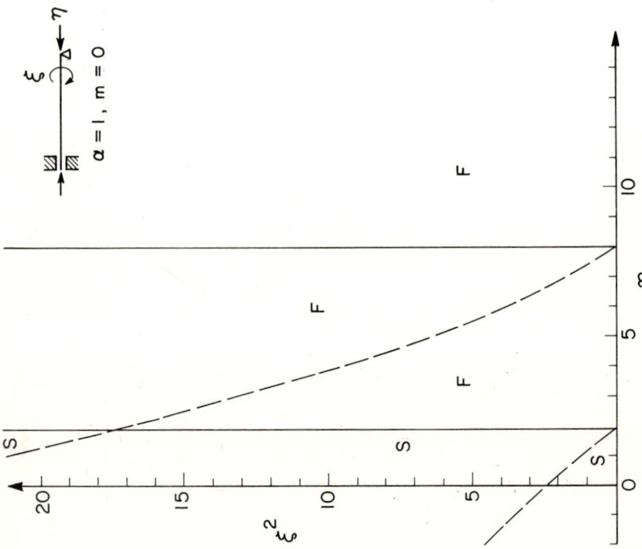

Figure 5.15

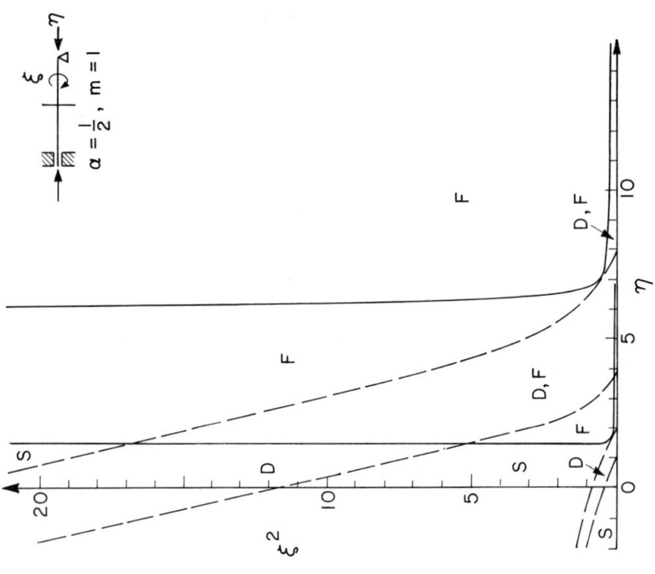

Figure 5.18

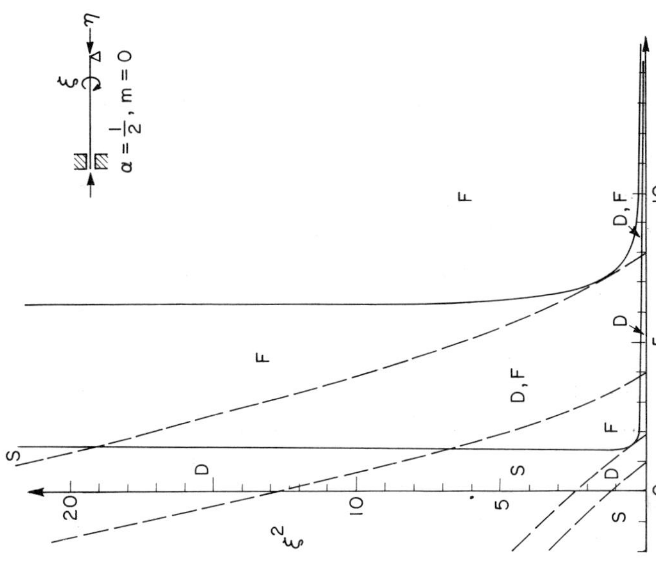

Figure 5.17

not a general one for all systems (5.91). The convexity properties of the flutter boundaries in Figures 5.13, 5.14, 5.17 and 5.18 also may not exist in all cases.

5.9 Example 3: Fluid-conveying pipes

Consider now a circular elastic pipe of mass per unit length m_p conveying a fluid of mass per unit length m_f at constant velocity v. Let x denote the axial coordinate, t the time, $y(x, t)$ the transverse displacement, $s = EI$ the elastic rigidity, and l the length of the pipe. Assuming small displacements, the equation of motion can be established [59, 68] as

$$sy'''' + m_f v^2 y'' + 2m_f v\dot{y}' + (m_f + m_p)\ddot{y} = 0. \qquad (5.125)$$

To derive this equation one observes that the x component of the velocity can be assumed to be v (small displacements) and the y component is

$$\dot{y} + y'v.$$

The kinetic energy in a length dx of the system is

$$dT = \{\tfrac{1}{2}m_p\dot{y}^2 + \tfrac{1}{2}m_f[v^2 + (\dot{y} + y'v)^2]\}\, dx.$$

The strain energy is given by

$$dU = \tfrac{1}{2}EI(y'')^2\, dx.$$

Using Hamilton's principle (2.45), one has

$$\delta \int_{t_1}^{t_2}\int_{x_1}^{x_2} \{\tfrac{1}{2}m_p\dot{y}^2 + \tfrac{1}{2}m_f[v^2 + (\dot{y} + y'v)^2] - \tfrac{1}{2}EI(y'')^2\}\, dx\, dt = 0.$$

Performing the variation and integrating by parts results in the differential equation (5.125).

Define the nondimensional quantities

$$\xi^2 = \frac{v^2 l^2 m_f}{\pi^2 EI}, \qquad \omega^2 = \frac{\bar{\omega}^2 l^4 (m_f + m_p)}{\pi^4 EI}, \qquad \gamma = \frac{m_f}{m_f + m_p}$$

where $\bar{\omega}$ is the frequency of vibration.

Using Galerkin's procedure with normalized modes of free vibration yields a system of equations of the form (5.21).

If both ends of the pipe are simply supported, one obtains

$$U = \begin{bmatrix} 1 & 0 \\ 0 & 16 \end{bmatrix}, \qquad F = \begin{bmatrix} 0 & -\rho \\ \rho & 0 \end{bmatrix}, \qquad E = \begin{bmatrix} 1 & 0 \\ 0 & 4 \end{bmatrix}, \qquad M = \begin{bmatrix} 1 & 0 \\ 0 & 1 \end{bmatrix},$$

with the use of the first two vibration modes as before, where $\rho = 256\,\gamma/9\pi^2$. This system belongs to Category (b) of Section 5.2. The characteristic curves for $\rho^2 = 0{\cdot}5$, $1{\cdot}0$ and $1{\cdot}5$ are shown in Figures 5.19(a), (b), and (c) respectively. The following sequences of behaviours are observed: stability, divergence, flutter, divergence in Figure 5.19(a); stability, divergence, stability, flutter in Figures 5.19(b) and (c). In each case the second interval of stability on the ξ^2 axis is quite small. The characteristic curves for $G = 0$ are again straight lines, and here they can be related to a simply-supported column under an axial load of magnitude $m_f v^2$. The upper bound for ω^2 in the first stability interval is a good approximation to its actual value in the gyroscopic system.

If one end of the pipe is clamped and the other is simply supported, a two-mode analysis leads to the characteristic curves shown in Figures 5.20(a), (b) and (c) for parameter values of $\bar{\rho}^2 = 0{\cdot}5$, $1{\cdot}0$, and $1{\cdot}5$ respectively, where $\bar{\rho}^2 = 2{\cdot}5\gamma$. This case belongs to Category (c) of Section 5.2. In Figure 5.20(a) the sequence of behaviour on the ξ^2 axis is stability, divergence; in Figure 5.20(b), stability, divergence, flutter; and in Figure 5.20(c), stability, divergence, stability, flutter. The curves for $G = 0$ are not straight lines here; the fundamental curve is concave toward the origin as expected and is close to the fundamental curve of the gyroscopic system in the first stability interval.

It is remarkable that, in contrast with rotating shafts, fluid-conveying pipes exhibit flutter without requiring the presence of other parameters such as an axial compression force. If axial force is also present, however, the stability behaviour of the pipe may be altered, and the effect of such a force is discussed in [69]. A review of the dynamic stability of fluid-conveying pipes is given by Paidoussis and Issid [70]. Other investigations include [71].

5.10 Dissipative gyroscopic systems

Introduction

It has been shown in the preceding sections that gyroscopic forces can stabilize an unstable conservative system, although a stable conservative system cannot be destabilized by such forces. What happens when dissipative forces are also considered? In the remaining sections of this chapter this question will be tackled. Two types of damping, namely *internal* and *external damping*, will be introduced and their effect on the behaviour of the system will be studied.

If dissipative forces with a positive definite damping matrix are added to the gyroscopic system (5.1), the Kelvin-Tait-Chetaev (KTC) theorem applies. This theorem provides a useful criterion of stability, and will be

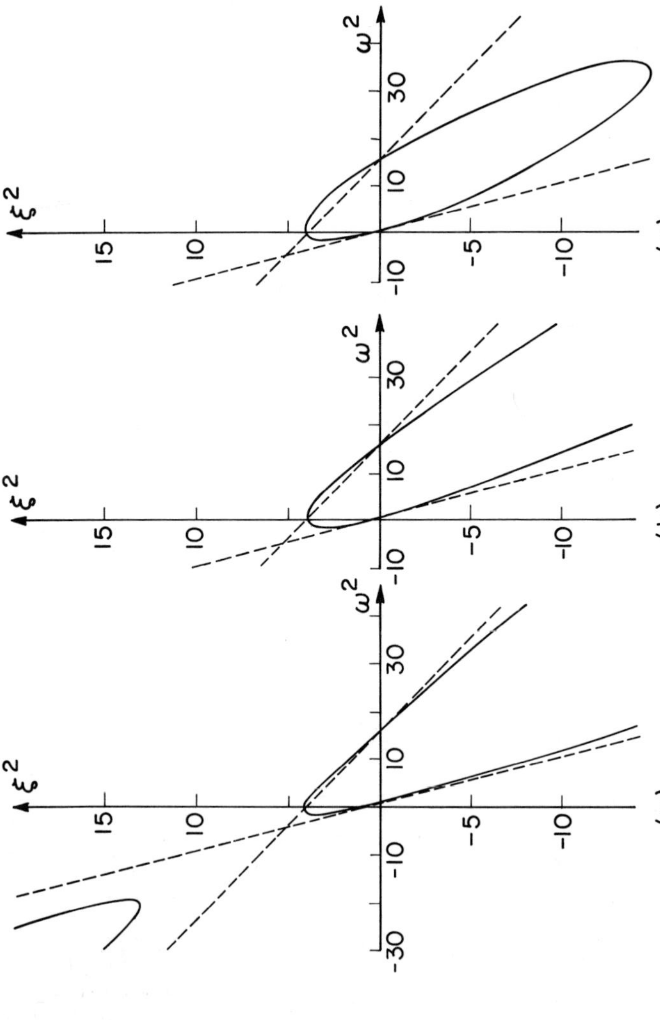

Figure 5.19

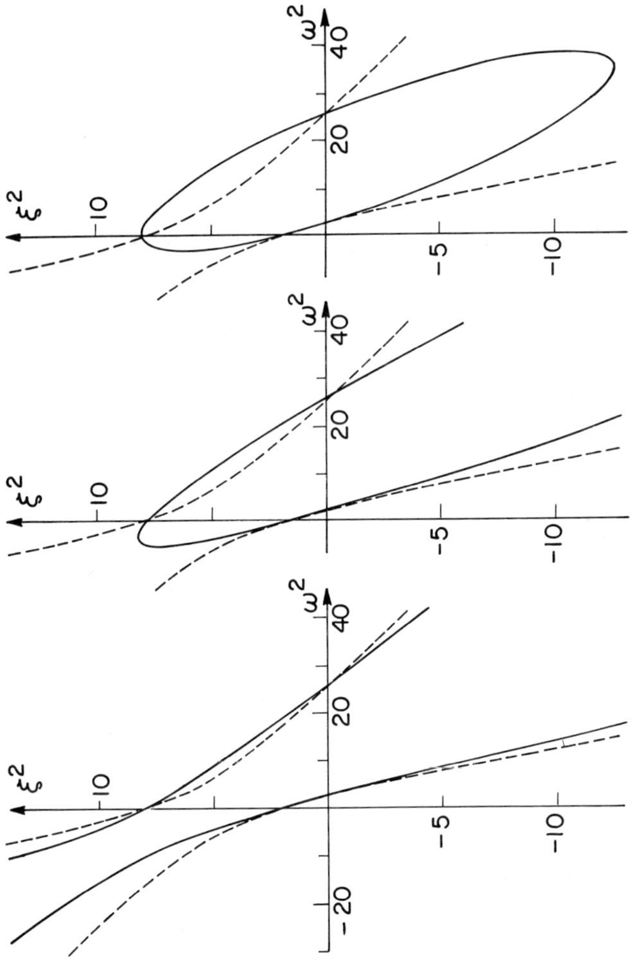

Figure 5.20

discussed in this section. Many physical systems (particularly in space mechanics), however, have a positive semi-definite damping matrix only, and inference of asymptotic stability via the KTC theorem is not directly possible. In attitude control systems, for example, energy is dissipated only by relative displacement of satellite parts, and the damping matrix is positive semi-definite rather than positive definite. Certain extensions and generalizations of the KTC theorem concerning this case have appeared in the last ten years.

Zajac [72, 73], recognizing the unnecessarily restrictive nature of the assumption concerning a positive definite damping matrix, introduced the notion of "pervasive-damping" associated with a positive semi-definite damping matrix D. In a pervasively damped system, any motion other than that corresponding to the null solution induces energy dissipation; this type of damping is, therefore, also referred to as "complete damping" which of course includes $D>0$. In other words, damping is pervasive if the dissipation function $R=\frac{1}{2}\langle\dot{q}, D\dot{q}\rangle$ is nonnegative and can be zero (for $t>0$) only when the system is at the equilibrium state (for $t>0$). Previous results for $D>0$, and in particular the KTC theorem, hold in the case of complete damping. Pringle [74, 75] generalized the KTC theorem in a different way. He constructed a function of generalized coordinates (potential) whose positive definiteness determines the stability of a gyroscopic system, and demonstrated asymptotic stability for complete damping. Roberson [76] proposed a method for determining whether or not a linear system is pervasively damped when $D\geq 0$. Connell [77] presented another method for the same purpose. Müller [78] discussed pervasive damping from the "observability" point of view.

Another type of damping, so called "constraint damping", was introduced in the last decade in the literature concerning space mechanics. Kane and Barba [79], investigating the stability of a dissipative gravity-oriented spinning satellite, obtained some results that did not fall within the scope of the KTC theorem and/or its extensions mentioned above. Likins [80, 81] observed that, in this case, the form of the linearized equations of motion is different. Actually, constraint damping is equivalent to *external damping* arising in the elastic shaft problems which were discussed in conjunction with internal damping by Dimentberg [82], Bolotin [83] and Ziegler [13]. Mingori [84] generalized the KTC theorem to cover the systems with constraint damping. The role of damping in vibration problems has been reviewed by Crandall [85].

Internal damping and the KTC theorem

If dissipative forces with a positive semi-definite damping matrix are added to the gyroscopic system (5.1), the equations of motion take the form

$$M\ddot{q}+(D+G)\dot{q}+Kq=0 \tag{5.126}$$

where $D \geq 0$.

According to the KTC theorem, an equilibrium position which is stable under purely potential forces (nongyroscopic nondissipative system or the system with $D=G=0$) remains stable with the addition of gyroscopic and dissipative forces. If $D>0$, the dissipative system is asymptotically stable. On the other hand, an equilibrium position which is unstable under potential forces remains unstable with the addition of gyroscopic and dissipative forces if the latter have complete dissipation.

One of the implications of this theorem is that an equilibrium state of the gyroscopic conservative system which was stabilized by gyroscopic forces will lose its stability with the addition of dissipation. In other words, all the higher regions of stability are wiped out by dissipation while the fundamental region of stability remains unaffected.

Considering again the function

$$H=\tfrac{1}{2}\langle \dot{q}, M\dot{q}\rangle+\tfrac{1}{2}\langle q, Kq\rangle$$

and its time derivative

$$\dot{H}=-\langle \dot{q}, D\dot{q}\rangle,$$

one observes that if $K>0$, H is a Lyapunov function and the dissipative system is stable if the nongyroscopic conservative system is stable. The former is asymptotically stable if, in addition to K, the damping matrix D is also positive definite.

To prove the second part of the theorem, Chetayev introduced another function and applied Lyapunov's instability theorem. Here, an alternative approach will be followed. Consider the eigenvalue problem

$$[\lambda^2 M+\lambda(D+G)+K]u=0 \tag{5.127}$$

which is obtained by introducing $q=u\mathrm{e}^{\lambda t}$ into (5.126). Pre-multiplying (5.127) by $\bar{u}'$ one obtains

$$\lambda^2+\lambda(d+ig)+k=0$$

where it is assumed that $\langle \bar{u}, Mu\rangle=1$; $d>0$ if $D>0$ and g and k are real.

Assuming further that

$$\lambda=\gamma+i\omega,$$

where γ and ω are real, results in the equations

$$\left.\begin{aligned}&\gamma^2-\omega^2+\gamma d-\omega g+k=0\\&2\gamma\omega+\gamma g+\omega d=0\end{aligned}\right\} \tag{5.128}$$

From the second equation one has

$$\omega = -\frac{\gamma g}{2\gamma + d}$$

which upon substitution in the first equation yields

$$\gamma^4 + a_1\gamma^3 + a_2\gamma^2 + a_3\gamma + a_4 = 0 \tag{5.129}$$

where

$$\left.\begin{aligned} &a_1 = 2d, \quad a_2 = \tfrac{1}{4}(5d^2 + g^2 + 4k) \\ &a_3 = \frac{d}{4}(d^2 + g^2 + 4k), \quad a_4 = \tfrac{1}{4}kd^2. \end{aligned}\right\} \tag{5.130}$$

The Routh-Hurwitz criterion (2.41) for this system is equivalent to requiring that

$$\begin{gathered} a_i > 0, \qquad i = 1, 2, 3, 4 \\ \text{and} \\ (a_1 a_2 - a_3)a_3 - a_1^2 a_4 > 0 \end{gathered} \tag{5.131}$$

for stability. If $d > 0$ and $k > 0$ for all eigenvectors normalized with respect to M, then (5.131) is satisfied and the system is asymptotically stable. If the conservative system with $D = G = 0$ is unstable so that for some vector(s) $k < 0$, the conditions (5.131) are violated and the system is unstable.

On the basis of (5.128) one can have an alternative stability criterion. Thus, expressing γ in terms of ω by using (5.128) the following can be stated:

A necessary and sufficient condition for stability is

$$\begin{gathered} \frac{\omega d}{2\omega + g} > 0 \qquad (\text{for} \quad 2\omega + g \neq 0) \\ \text{or} \\ d > 0 \quad \text{and} \quad k - \omega g - \omega^2 > 0 \end{gathered} \tag{5.132}$$

for all eigenvectors normalized with respect to M.

In analogy with conservative systems (Chapter 3), the positive definiteness is not a necessary condition for asymptotic stability of a gyroscopic system. If $D \geq 0$, but the system is *pervasively damped* so that none of the eigenvectors of the undamped gyroscopic system lie in the null space of D, then the system (5.126) is asymptotically stable. This is a necessary and sufficient condition which follows readily from the stability conditions

(5.131) since by definition $d>0$ for all eigenvectors in the case of pervasive damping. Obviously, pervasive damping is also necessary for asymptotic stability. An alternative proof can be given by following the procedure of Section 3.5.

Internal and external damping

Consider now the more general case which involves both internal and external damping. The equations of motion are given by (2.64) and will be expressed here as

$$M\ddot{q}+(D_i+D_e+G)\dot{q}+(K+S)q=0 \tag{5.133}$$

where D_i (internal damping) is used for the sake of symmetry in appearance. Here, $D_i\geq 0$, $D_e\geq 0$ and $S=-S'$ while M, G, and K are as defined before. The associated eigenvalue problem is

$$[\lambda^2 M+\lambda(D_i+D_e+G)+K+S]u=0. \tag{5.134}$$

Pre-multiplying (5.134) by $\bar{u}'$ yields

$$\lambda^2+\lambda(d_i+d_e+ig)+k+is=0 \tag{5.135}$$

where $\langle\bar{u}, Mu\rangle=1$, $d_i=\langle u, D_i u\rangle\geq 0$, $d_e=\langle\bar{u}, D_e u\rangle\geq 0$ and k, s, and g are real.

Introducing $\lambda=\gamma+i\omega$ into (5.135) yields

$$\begin{aligned}&\gamma^2-\omega^2+\gamma d-\omega g+k=0\\&2\gamma\omega+\gamma g+\omega d+s=0\end{aligned} \tag{5.136}$$

where $d=d_i+d_e$.

Substituting for ω from the second equation into the first one results in an equation of the form (5.129) with

$$\begin{aligned}&a_1=2d, \quad a_2=\tfrac{1}{4}(5d^2+g^2+4k)\\&a_3=\frac{d}{4}(d^2+g^2+4k), \quad a_4=\tfrac{1}{4}(kd^2+sdg-s^2)\end{aligned} \tag{5.137}$$

Remarkably, only the fourth coefficient a_4 is different from the values given in (5.130). On the basis of (5.131), the following stability criterion can be established:

The system is stable (asymptotically) if and only if

$$d>0 \quad \text{and} \quad kd^2+sdg-s^2>0 \tag{5.138}$$

for all eigenvectors u normalized with respect to M. A similar result was obtained in [53] for continuous systems.

An alternative criterion is obtained by expressing γ in terms of ω. Thus, from (5.136) it follows that the system is stable if and only if

$$\frac{\omega d+s}{2\omega+g}>0 \quad (\text{for} \quad 2\omega+g\neq 0) \tag{5.139}$$

or

$$d>0 \quad \text{and} \quad k-\omega g-\omega^2>0. \tag{5.140}$$

If $2\omega+g=0$ then $\omega d+s=0$ and substituting $s=-\omega d$ in (5.138) shows that (5.138) is identical to (5.140).

It is noted that (5.138) is quite general and readily reduces to the case in which external damping is absent. The stability of systems discussed so far can also be based on this criterion. For undamped systems ($d=0$), however, modified conditions of stability must be derived on the basis of (5.136), since asymptotic stability is ruled out in this case. It is not difficult to deduce from (5.136) that stability (not asymptotic) is ensured if

$$g^2+4k>0 \quad \text{and} \quad s=0 \qquad (k\neq 0)$$

or (5.141)

$$-\omega^2-\omega g+k=0 \quad \text{and} \quad s=0 \qquad (k\neq 0)$$

for all eigenvectors normalized with respect to M.

Assuming now that s is associated with circulatory forces, an important result is immediately noted: *A nondissipative gyroscopic system is in general unstable if circulatory forces are also present [52, 53].*

Returning to the general stability criterion (5.138) one can state the criticality and instability conditions for $d>0$ as follows:

$$\text{The equilibrium is} \begin{cases} \text{critical} \\ \text{unstable} \end{cases} \text{if and only if} \quad \begin{matrix} kd^2+sdg-s^2=0 \\ kd^2+sdg-s^2<0 \end{matrix} \tag{5.142}$$

for some mode(s).
Similar criteria can be stated on the basis of (5.140) and (5.139).

An important interpretation of (5.138) is concerned with the effect of external damping on the stability of an internally damped gyroscopic system. Writing (5.138) in the form

$$d>0, \qquad kd^2+\left(\frac{dg}{2}\right)^2-\left(\frac{dg}{2}-s\right)^2>0,$$

one deduces that a gyroscopic system which is unstable under internal damping may be stabilized with the addition of external damping. In other words, the destabilizing effect of internal damping on the gyroscopically stabilized regions can be cancelled by external damping. In fact, experiments show that a gyroscopic system is often stable when it should be unstable according to the KTC theorem. There must, therefore, exist additional influences which suspend the destabilizing effect of internal damping. One of these influences is clearly external damping. Dimentberg [82] discusses other effects. The combined effect of internal and external damping on rotating systems will be discussed in the following sections.

5.11 Effect of dissipation on the stability boundaries of rotating systems

Consider a rotating system with external and internal damping. In Section 5.6, equations of motion of an undamped rotating system were derived. If in addition to forces considered there, external and internal damping forces are taken into account, one has

$$F_e = -md_e\dot{x}e_x - md_e\dot{y}e_y,$$

resulting from damping of motion with respect to stationary coordinates, and

$$F_i = -md_i(\dot{x} + \xi y)e_x - md_i(\dot{y} - \xi x)e_y$$

resulting from damping of motion with respect to rotating coordinates. Here d_e and d_i are the coefficients of external and internal damping respectively, and they are multiplied by m for convenience.

In rotating coordinates, (F_e) and (F_i) take the form

$$\left.\begin{aligned} &F_e = -md_e(\dot{p} - \xi q)e_p - md_e(\dot{q} + \xi p)e_q \\ &\text{and} \\ &F_i = -md_i\dot{p}e_p - md_i\dot{q}e_q. \end{aligned}\right\} \tag{5.143}$$

Applying Newton's law to a particle of mass m and generalizing it to n particles yields the linear differential equations

$$\begin{aligned} &M\ddot{p} + (d_i + d_e)M\dot{p} - 2\xi M\dot{q} + (U - \eta E - \xi^2 M)p - d_e\xi Mq = 0 \\ &M\ddot{q} + (d_i + d_e)M\dot{q} + 2\xi M\dot{p} + (V - \eta E - \xi^2 M)q + d_e\xi Mp = 0 \end{aligned} \tag{5.144}$$

in the rotating frame, where an axial force η is also included and the matrices are as defined before.

The characteristic equation is

$$\Delta = \begin{vmatrix} \Delta_{11} & \Delta_{12} \\ \Delta_{21} & \Delta_{22} \end{vmatrix} = 0 \tag{5.145}$$

where

$$\Delta_{11} = -\omega^2 M + i\omega(d_i + d_e)M + U - \eta E - \xi^2 M$$
$$\Delta_{12} = -\Delta_{21} = -2i\omega\xi M - d_e\xi M$$
$$\Delta_{22} = -\omega^2 M + i\omega(d_i + d_e)M + V - \eta E - \xi^2 M.$$

For the nondissipative system ($d_i = d_e = 0$), Δ is a function of ω^2, η and ξ^2, and the divergence and flutter boundaries are defined by $\omega^2 = 0$, $\Delta = 0$ and by $\partial\Delta/\partial\omega^2 = 0$, $\Delta = 0$, respectively, as discussed earlier. In the presence of damping, the divergence boundary is again defined by the same equations, but the double-root condition $\partial\Delta/\partial\omega^2 = 0$ for the flutter boundary is generally replaced by the condition that ω is real, although the former condition may still yield a portion of the flutter boundary which lies totally in an unstable region. The stability boundary of the system in the $(\xi^2 - \eta)$ plane consists of divergence and flutter boundaries surrounding stable regions. Two cases will again be considered separately:

(i) *Uniform system*: $U = V$

In this case, introducing the complex displacement $r = p + iq$, (5.144) can be written as

$$M\ddot{r} + (d_i + d_e)M\dot{r} + 2i\xi M\dot{r} + (U - \eta E - \xi^2 M)r + id_e\xi Mr = 0. \qquad (5.146)$$

Assuming $r = ue^{i\omega t}$ and pre-multiplying by $\bar{u}'$ yields

$$-\omega^2 m + i(d_i + d_e)\omega m - 2\omega\xi m + (a - \eta e - \xi^2 m) + id_e\xi m = 0 \qquad (5.147)$$

which at the onset of flutter (ω real) results in

$$-\omega^2 m - 2\omega\xi m + a - \eta e - \xi^2 m = 0$$

and (5.148)

$$\omega(d_i + d_e)m + d_e\xi m = 0$$

where $m = \langle\bar{u}, Mu\rangle$, $a = \langle\bar{u}, Uu\rangle$, etc. as before.

Substituting for ω from the second equation into the first, one has

$$a - \eta e - \left(\frac{d_i}{d_i + d_e}\right)^2 \xi^2 m = 0 \qquad (5.149)$$

which implies

$$\left|U - \eta E - \left(\frac{d_i}{d_i + d_e}\right)^2 \xi^2 M\right| = 0 \qquad (5.150)$$

since (5.149) is positive for sufficiently small η and ξ, and vanishes at the onset of flutter.

(5.150) defines the flutter boundary. Comparing with the divergence boundary of nondissipative systems and recalling Theorem 5.3, it is seen that the presence of $(d_i/(d_i+d_e))^2$ does not alter the essential properties of the boundary here; hence the following theorem can be stated [86]:

THEOREM 5.8. *The initial flutter boundary of a uniform rotating system cannot have convexity toward the origin of the* (ξ^2, η) *plane.*

COROLLARY 5.1. *If the matrices U, E and M can be diagonalized simultaneously, the flutter boundary is in the form of straight lines.*

The divergence boundary of the system does not exist or rather consists of the critical points on the η axis given by $|U-\eta E|=0$ provided $d_e \neq 0$. This can readily be deduced from (5.148) which indicates that for $\omega=0$, $\xi=0$ if $d_e \neq 0$. If $d_e=0$, the KTC theorem applies and the divergence boundary (which is also the stability boundary) is given by the divergence boundary (5.102) of the undamped system. On the other hand, if $d_i=0$, the flutter boundary (5.150) takes the form of straight lines $(\eta=\eta_j)$ parallel to the ξ^2 axis where η_j is defined by

$$|U-\eta E|=0.$$

These lines, however, were shown to be the flutter boundary (5.97) of the undamped system and in view of (5.150) one has the following result [86]:

THEOREM 5.9. *The flutter (divergence) boundary of the undamped system provides an upper (lower) bound for the flutter boundary of the damped system as* d_i *and* d_e *are varied.*

An increase in d_e enlarges the region of stability, effectively rotating the flutter boundary clockwise around the critical point on the η axis. An increase in d_i, on the other hand, reduces the region of stability (see Example 4).

(ii) *Asymmetric system:* $U \neq V$

It follows from (5.145) that the divergence boundary is independent of d_i and is given by

$$\begin{vmatrix} U-\eta E-\xi^2 M & -d_e\xi M \\ d_e\xi M & V-\eta E-\xi^2 M \end{vmatrix}=0. \tag{5.151}$$

Consider the first stability region of the undamped $(d_e=0)$ system; since the matrices on the main diagonal of (5.151) are then positive definite and the off-diagonal matrices constitute a skew-symmetric matrix, (5.151) cannot hold in the primary stability region (Section 1.3, Exercise 7). Hence the following result can be formulated:

THEOREM 5.10. *The initial divergence boundary of the undamped system is a lower bound to the divergence boundary of the damped system.*

An alternative proof is given in [86]. As will be demonstrated by the examples, the divergence boundary of a two-degree-of-freedom system takes the form of a curve between the two divergence boundaries of the undamped system, the shape of which and the size of the instability region change with d_e. The instability region shrinks and tends to zero for sufficiently large d_e.

To obtain general results concerning the flutter boundary of an asymmetric system is harder, and as in the case of undamped systems, attention will be restricted here to situations in which M, U, V and E are simultaneously diagonalizable. In practice, U is a scalar multiple of V, and such a transformation is possible, for example, in the case of simply supported shafts. The characteristic equation (5.145) is then uncoupled, and one has the product of 2×2 determinants.

Considering the first region and assuming that $M=I$ for simplicity, the expansion of (5.145) yields

$$(-\omega^2+k_1)(-\omega^2+k_2)-4\omega^2\xi^2-\omega^2(d_i+d_e)^2+d_e\xi^2 \\ +i\omega[(-2\omega^2+k_1+k_2)(d_i+d_e)+4d_e\xi^2]=0 \quad (5.152)$$

where

$$k_1=u-\eta e-\xi^2, \qquad k_2=v-\eta e-\xi^2,$$

and u, v and e are the diagonal elements of the corresponding matrices. At the onset of flutter ω is real, and equating the real part of (5.152) to zero results in an equation in ω^2. If both d_i and d_e are small, the roots (ω^2) of this equation can only be real if

$$(k_1+k_2+4\xi^2)^2-4k_1k_2>0. \quad (5.153)$$

But this inequality is precisely the condition for the undamped system to have real frequencies ω^2. It follows that at the flutter boundary of the damped system, the undamped system has real frequencies ω^2. Thus, the following theorem can be stated:

THEOREM 5.11. *The flutter boundary of the undamped system is an upper bound for the flutter boundary of the damped system.*

If $d_e = 0$, the KTC theorem applies and the boundary reduces to $k_1 = 0$, $k_2 = 0$, the initial boundary being associated with k_1 if $u < v$. If $d_i = 0$, it can be shown that the flutter boundary lies between the flutter boundary of the undamped system and its vertical asymptote $\eta = (u+v)/2e$ which is normally a very narrow region. To show this, equate the imaginary part of (5.152) to zero (note that $\omega = 0$ leads to divergence and is not used), substitute for ω^2 in the real part and solve for ξ^2 or η to get

$$\xi^2 = \frac{(u-v)^2}{8(2\eta e - u - v)} - \frac{d_e^2}{4} \tag{5.154a}$$

or

$$\eta = \frac{u+v}{2e} + \frac{(u-v)^2}{4(4\xi^2 + d_e^2)} \tag{5.154b}$$

respectively. Comparing with (5.105) and (5.107), one observes that $d_e = 0$ leads to the flutter boundary of the undamped system, while increasing d_e draws the boundary (5.154) closer to its asymptotes. This is interesting because it amounts to a reduction of the stability region. The addition of a small internal damping shifts the flutter boundary to the left of the vertical asymptote, and the region of instability increases with d_i.

5.12 Example 4: Dissipative rotating flexible shafts

Vibrations and stability of dissipative rotating shafts have been studied by a number of authors [e.g., 86–93, 82, 83, 13]. Here, divergence and flutter boundaries of dissipative shafts will be examined.

Consider again the flexible shaft which was analyzed in Section 5.8, and suppose both internal and external damping are present. The former is assumed to be given by $\mu \bar{d}_i(\dot{y}_1, \dot{y}_2)$ which is related to transverse velocities in the rotating frame, and the latter which is given by μd_e times velocities relative to stationary coordinates is expressed as $\mu \bar{d}_e[(\dot{y}_1 - \bar{\xi} y_2), (\dot{y}_2 + \bar{\xi} y_1)]$ in the rotating frame. Adding these damping forces to those acting on the element $d\varphi$ in Figure 5.9 leads to the differential equations

$$\begin{aligned}
&\mu \ddot{y}_1 + (\bar{d}_i + \bar{d}_e)\mu \dot{y}_1 - 2\bar{\xi}\mu \dot{y}_2 + s_1 y_1'''' + \bar{\eta} y_1'' - \bar{\xi}^2 \mu y_1 - \bar{d}_e \mu \bar{\xi} y_2 = 0 \\
&\mu \ddot{y}_2 + (\bar{d}_i + \bar{d}_e)\mu \dot{y}_2 + 2\bar{\xi}\mu \dot{y}_1 + s_2 y_2'''' + \bar{\eta} y_2'' - \bar{\xi}^2 \mu y_2 + \bar{d}_e \mu \bar{\xi} y_1 = 0
\end{aligned} \tag{5.155}$$

with the boundary conditions (5.118).

Assuming motion of the form (5.119) results in

$$-\bar{\omega}^2\mu z_1+i\bar{\omega}(\bar{d}_i+\bar{d}_e)\mu z_1-2i\bar{\omega}\bar{\xi}\mu z_2+s_1 z_1''''+\bar{\eta}z_1''-\bar{\xi}^2\mu z_1-\bar{d}_e\mu\bar{\xi}z_2=0$$
$$-\bar{\omega}^2\mu z_2+i\bar{\omega}(\bar{d}_i+\bar{d}_e)\mu z_2+2i\bar{\omega}\bar{\xi}\mu z_1+s_2 z_2''''+\bar{\eta}z_2''-\bar{\xi}^2\mu z_2+\bar{d}_e\mu\bar{\xi}z_1=0 \tag{5.156}$$

Introducing the solutions (5.121) and following the procedure as in Example 2 leads to the matrix equation

$$-\omega^2\begin{bmatrix} M & 0 \\ 0 & M\end{bmatrix}+i\omega(d_i+d_e)\begin{bmatrix} M & 0 \\ 0 & M\end{bmatrix}+2i\omega\xi\begin{bmatrix} 0 & -M \\ M & 0\end{bmatrix}$$
$$+\begin{bmatrix} U-\eta E-\xi^2 M & 0 \\ 0 & V-\eta E-\xi^2 M\end{bmatrix}+d_e\xi\begin{bmatrix} 0 & -M \\ M & 0\end{bmatrix}=0 \tag{5.157}$$

where the matrices *M*, *U*, *V* and *E* are given by (5.123) and the nondimensional quantities (5.122) together with

$$d_i=\bar{d}_i\frac{l^2}{\pi^2}\sqrt{\frac{\mu}{s_1}},\qquad d_e=\bar{d}_e\frac{l^2}{\pi^2}\sqrt{\frac{\mu}{s_1}}$$

are used.

If both ends of the shaft are simply supported, the use of the sine series results in the matrices (5.124). In the case of the uniform shaft ($\alpha=1$), the flutter boundary is determined from (5.150) and is shown in Figure 5.21. Regions of stability, divergence and flutter are denoted by *S*, *D* and *F* respectively. Similarly, *D* and *F* are used on the curves to indicate divergence and flutter boundaries. Full (dashed) lines belong to

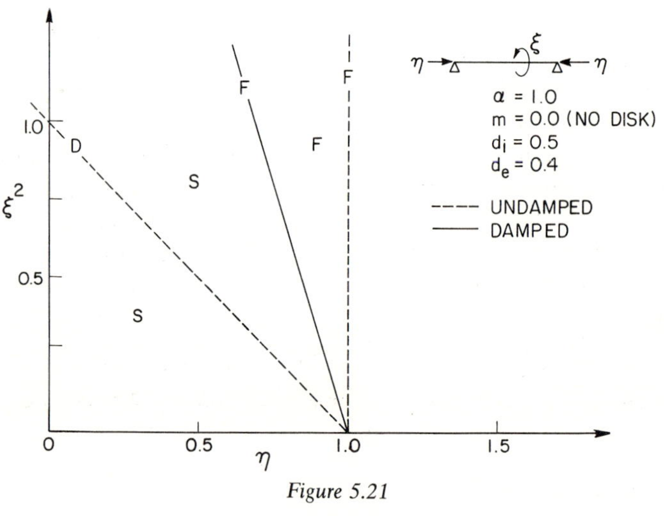

Figure 5.21

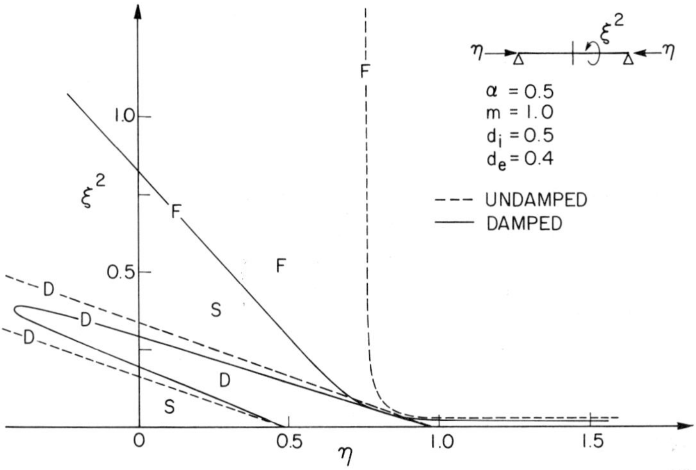

Figure 5.22

damped (undamped) systems, enabling comparisons. The uniform shaft remains stable until flutter occurs. In this case, the flutter boundary is a straight line (Corollary 5.1), and rotates around the point on the η axis between the two boundaries of the undamped system as d_i and d_e are varied. For $d_e = 0$ ($d_i = 0$) it coincides with the divergence (flutter) boundary of the undamped system.

Figures 22–24 show the boundaries for an asymmetric shaft. The divergence boundary lies between the two straight boundaries of the

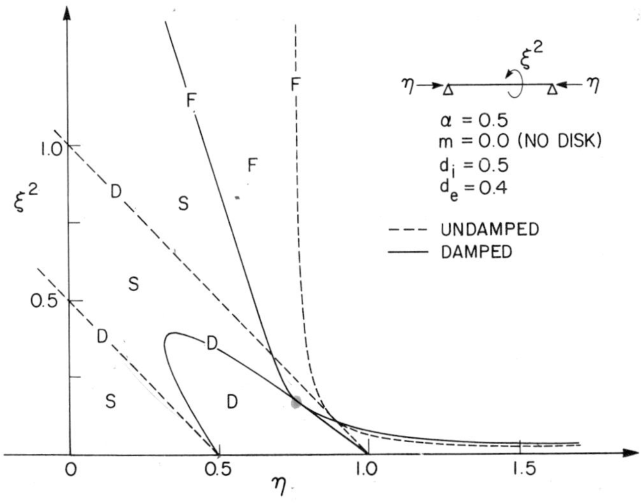

Figure 5.23

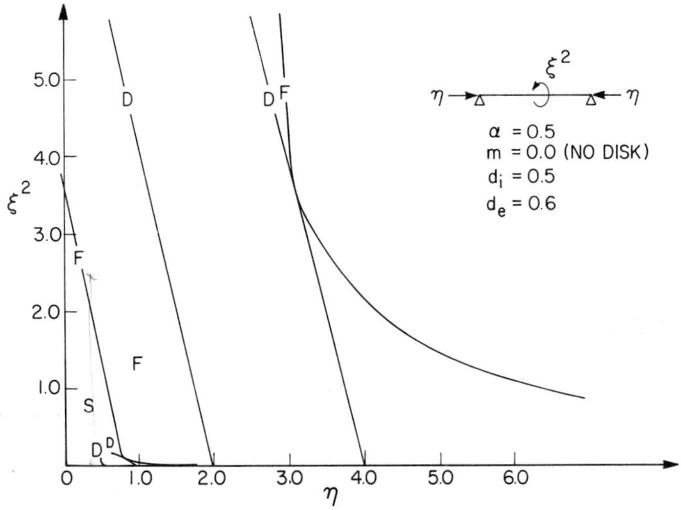

Figure 5.24

undamped system, and the divergence region is reduced as d_e is increased. At one stage, intersection with the ξ^2 axis does not occur, resulting in stable values for ξ^2 until flutter occurs. This is a phenomenon observed before, but the two parameter analysis here sheds light on the mechanism leading to such phenomena. Removing the disk has a stabilizing effect for both the damped and undamped systems (Figure 5.23). As d_i tends to zero, the flutter boundary approaches the vertical asymptote of the undamped system and for $d_i = 0$ remains contained within the region surrounded by the flutter boundary and its asymptote as shown in the theory. Including higher order modes in the analysis does not have an effect on the boundaries (Figure 5.24). The lower portion of the flutter boundary (Figure 5.23) is obtained on the basis of a double-root, and lics totally in an unstable region.

Different boundary conditions do not alter the qualitative properties of divergence and flutter boundaries [92].

Problems

1. Introduce internal and external damping to the model considered in Example 1, determine divergence and flutter boundaries and show that the behaviour of the system is qualitatively similar to that of the flexible shafts analyzed in Example 4.

2. Introduce appropriate admissible functions ϕ_k to analyze a flexible shaft with one end simply supported and the other end clamped, under the influence of internal and external damping [103].

3. Considering the differential equation of the pipe analyzed in Section 5.9 which is also subjected to an axial compression force, determine the divergence and flutter boundaries and discuss the effect of the axial force on the behaviour of the system.

4. Examine the behaviour of a gyroscopic system under the influence of follower forces by appropriately interpreting and modifying the external damping terms considered in this chapter.

6

Circulatory Systems

In this chapter, general nonconservative systems associated with circulatory forces will be examined. The distinction between this class of systems and pseudo-conservative systems (Chapter 4) is that the latter were assumed to be incapable of exhibiting flutter instability due to the weak asymmetry of matrices associated with circulatory forces. Systems which can exhibit flutter are called circulatory here following Ziegler [13] although in his terminology pseudo-conservative systems are not distinguished.

Circulatory systems have been studied by a number of authors. In 1967, Herrmann [94] reviewed the vibration and stability of elastic systems subjected to follower forces. A more recent review is presented by Sundararajan [95]. Various text books and monographs [13–15, 83] provide excellent background material concerning the behaviour of circulatory systems. Recent survey papers on aeroelasticity [96] and thermoelastic stability [97] are also concerned with related areas.

6.1 Properties of circulatory systems

As in Chapter 4, consider again the differential equation

$$M\ddot{q}+(U-\eta_k E^k)\,q=0 \tag{6.1}$$

or

$$M\ddot{q}+Kq=0 \tag{6.2}$$

where $k=1,2,\ldots,m$, and some E^k (hence K) is asymmetric while $M=M'>0$.

In general $M^{-1}K$ is not symmetrizable and can have complex eigenvalues for a given set of η_k. Assuming solutions of the form $q=u\mathrm{e}^{i\omega t}$, one has

$$(-\omega^2 M+K)\,u=0 \tag{6.3}$$

and

$$v'(-\omega^2 M+K)=0 \tag{6.4}$$

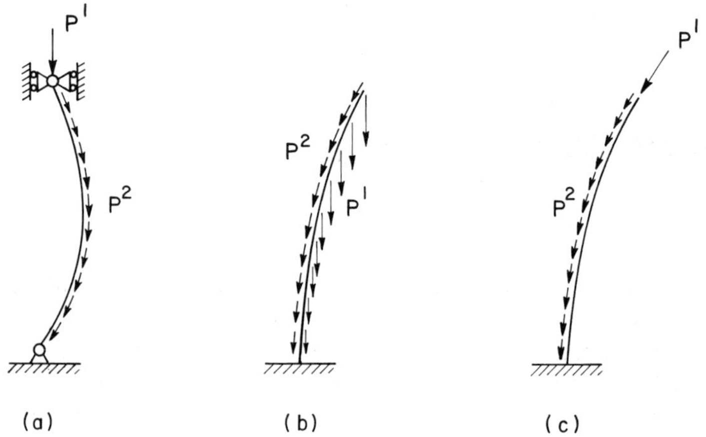

Figure 6.1

where v is the left eigenvector associated with the eigenvalue ω^2. The characteristic equation

$$\Delta(\omega^2, \eta_k) \equiv |-\omega^2 M + U - \eta_k E^k| = 0 \tag{6.5}$$

describes certain m dimensional characteristic surfaces in the $m+1$ dimensional $\omega^2 - \eta_k$ space. As one follows an arbitrary path emanating from the origin of the parameter space, the initially stable system may lose stability by divergence or by flutter, according to whether one root ω^2 of (6.5) vanishes or two roots coalesce, respectively. The locus of the points in the parameter space corresponding to instability by divergence is called the *divergence boundary*, while the points corresponding to flutter instability comprise the *flutter boundary* as before. The points on these two boundaries which are associated with initial loss of stability form the *stability boundary* of the system. Some typical examples are shown in Figure 6.1 whose boundaries are depicted in Figures 6.2–6.4. The divergence boundary is denoted by DB, the flutter boundary by FB, and the stability boundary is shaded. Example (a) is, of course, a pseudo-conservative system, and it is included here for completeness and comparison. Typical characteristic curves which may arise are shown in Figure 6.5.

It was shown in Section 2.5 that the flutter boundary is obtained by solving (6.5) with

$$\frac{\partial \Delta}{\partial \omega^2} = 0, \tag{6.6}$$

while the divergence boundary is defined by

$$|U - \eta_k E^k| = 0.$$

It immediately follows that *if* $\eta_k E^k$ *is skew-symmetric divergence instability is ruled* out since the above determinant cannot then vanish (Exercise 7, Section 1.3).

In general, for $\eta_k = 0$ the system (6.3) has real positive roots ω^2 and the system is stable. For $\eta_k \neq 0$, the roots ω^2 remain real until a pair of roots approaches each other and coalesces, beyond which point the pair becomes complex conjugates and flutter instability occurs. This is, however, not necessary for instability. In fact, the system may be unstable even when all the roots ω^2 are real and distinct but at least one $\omega^2 \leq 0$ (divergence instability). A comprehensive stability condition, embracing both divergence and flutter, can be obtained from the general criteria given in Section 5.10. Thus, on the basis of (5.141), one observes that the system (6.3) is stable if and only if

$$k > 0 \quad \text{and} \quad s = 0 \tag{6.7}$$

for all eigenvectors u normalized with respect to M so that $\langle \bar{u}, Mu \rangle = 1$, $\langle \bar{u}, Ku \rangle = k + is$, k being associated with the symmetric part of K and s with the skew-symmetric part of K. The condition (6.7) also follows from (6.3) directly upon pre-multiplying this equation by $\bar{u}'$. (6.7) implies that

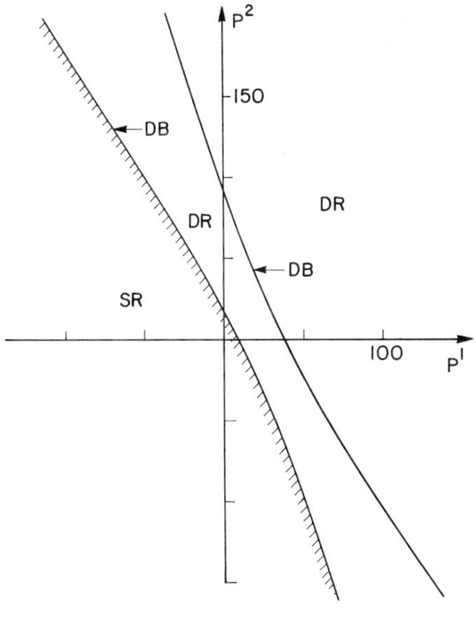

Figure 6.2

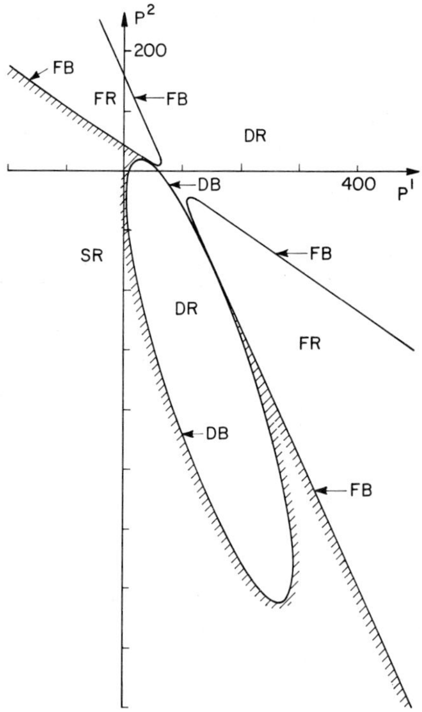

Figure 6.3

the eigenvectors are real and one, therefore, can replace the quantities in (6.7) by

$$k = \langle u, Ku \rangle$$
$$1 = \langle u, Mu \rangle \tag{6.8}$$

without loss of generality.

It is further noted that the matrix $M^{-1}K$ is symmetrizable as long as the roots ω^2 are real and distinct. This implies that the right (and left) eigenvectors are linearly independent before flutter occurs, and the matrix pencil $-\omega^2 M + K$ is *simple* in this region. It follows immediately from the definition of symmetrizable matrices (see also Chapter 4) and the above stability considerations that the following theorem can be formulated:

THEOREM 6.1. *A circulatory system is stable if and only if there exists a symmetric and positive definite matrix S_1 such that $S_1 M^{-1} K$ is symmetric and positive definite.*

COROLLARY 6.1. *If the matrix $S_1M^{-1}K$ is symmetric, flutter instability is ruled out.*

Pre-multiplying (6.3) by M^{-1} and S_1, it is clearly seen that

$$(-\omega^2 S_1 + S_1 M^{-1} K)u = 0 \tag{6.9}$$

yields real positive roots ω^2 under the conditions of Theorem 6.1, and real roots if Corollary 6.1 applies [44, 49].

At a flutter point where two roots ω_1^2 and ω_2^2 coalesce, the matrix $M^{-1}K$ is no longer symmetrizable and the eigenvectors are not linearly independent. At this point, the matrix pencil $-\omega^2 M + K$ is *defective* [98, 99]. An alternative flutter condition can be obtained by differentiating (6.3) with respect to ω^2 and using (6.6) which is equivalent to $\partial\eta_k/\partial\omega^2 = 0$

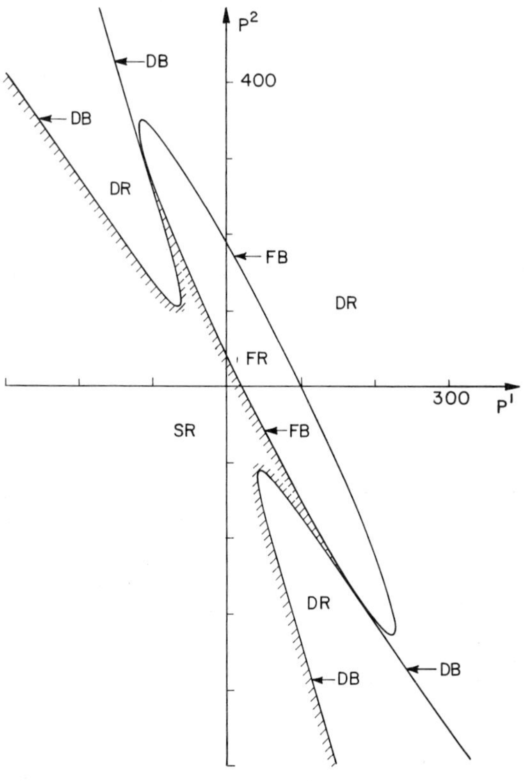

Figure 6.4

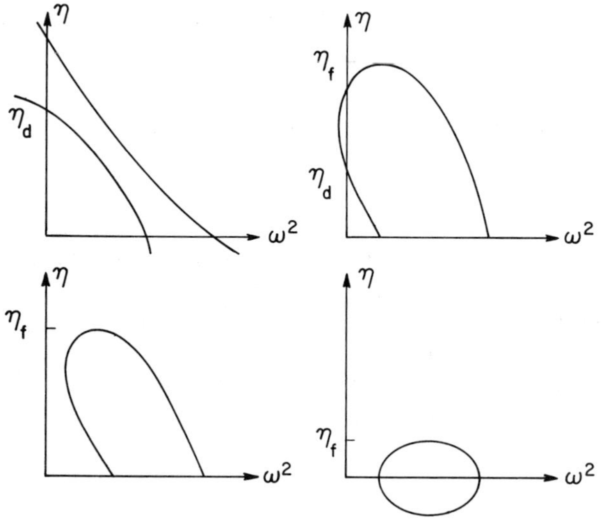

Figure 6.5

as was shown in Section 2.5. Thus, the differentiation yields

$$\left(-M+\frac{\partial \eta_k}{\partial \omega^2}E^k\right)u+(-\omega^2 M+U-\eta_k E^k)\frac{\partial u}{\partial \omega^2}=0 \tag{6.10}$$

which upon pre-multiplying by the left eigenvector v' results in the condition

$$\langle v, Mu\rangle = 0 \tag{6.11}$$

at flutter. It was shown by Lancaster [3] that in general a matrix pencil in the form of (6.3) is defective if (6.11) is satisfied. This is significant since the mathematical theory related to matrix pencils has been developed mostly with regard to simple (non-defective) pencils.

In the region associated with a symmetrizable matrix $M^{-1}K$, Rayleigh's quotient is available, and the real frequencies ω^2 are obtained as stationary values of this quotient. In general, however, the eigenvalue problem (6.3) is associated with the *generalized Rayleigh quotient* (Section 1.5)

$$R(x, y)=\frac{\langle y, Kx\rangle}{\langle y, Mx\rangle} \tag{6.12}$$

which is not restricted to the stable region. Normally, even when some frequencies ω^2 are complex, the matrix pencil $-\omega^2 M+K$ is simple although not symmetrizable (excluding the critical flutter points), and the frequencies of the system are given as the stationary values of (6.12),

$$\omega_j^2 = \frac{\langle v^j, Ku^j \rangle}{\langle v^j, Mu^j \rangle}, \qquad j = 1, 2, \ldots, n. \tag{6.13}$$

It was shown in Section 1.5 that if the matrix pencil is simple so that it has n linearly independent eigenvectors, the u^i and v^j can be chosen such that

$$\langle v^j, Mu^i \rangle = \delta_{ij}; \qquad i, j = 1, 2, \ldots, n. \tag{6.14}$$

The stationary values of the quotient $R(x, y)$ are then obtained as

$$\omega_j^2 = R(u^j, v^j) = \langle v^j, Ku^j \rangle. \tag{6.15}$$

6.2 Flutter instability and the extremum properties of the generalized Rayleigh quotient

In a vibration analysis, ω^2 is the eigenvalue, and at a flutter point the generalized Rayleigh quotient is not simple. In a direct stability analysis, however, one seeks the critical flutter values of the parameters, and the eigenvalue problem can be modified appropriately such that the defective pencil is avoided.

In this section it will be assumed that the system has one parameter η only (or equivalently, all the independent parameters are represented by one parameter). In a stability analysis, η can be treated as an eigenvalue by considering the quotient

$$R(x, y, \omega^2) = \frac{\langle y, (U - \omega^2 M)x \rangle}{\langle y, Ex \rangle} \tag{6.16}$$

where E corresponds to η.

Since the attention will be focussed on flutter instability, only pairs of vectors x, y which satisfy the flutter condition

$$\langle y, Mx \rangle = 0 \tag{6.17}$$

will be considered in what follows. The generalized Rayleigh quotient then reduces to

$$R(x, y) = \frac{\langle y, Ux \rangle}{\langle y, Ex \rangle}. \tag{6.18}$$

Consider a critical flutter point at which $\omega^2 = \omega_f^2$, $\eta = \eta_1$, and the eigenvectors u^1 and v^1 satisfy condition (6.11). At this fixed frequency,

(6.3) and (6.4) become

$$(-\omega_f^2 M + U - \eta E)u = 0$$
$$v'(-\omega_f^2 M + U - \eta E) = 0 \tag{6.19}$$

while the bi-orthogonality conditions (note that with η as the eigenvalue the repeated root at flutter is avoided) are given by

$$\langle v^j, Eu^i \rangle = \delta_{ij}, \qquad i, j = 1, 2, \ldots, n. \tag{6.20}$$

With the use of (6.11), (6.19) and (6.20) one can obtain the following:

$$\langle v^1, Uu^1 \rangle = \eta_1,$$
$$\langle v^s, Uu^s \rangle = \eta_s + \omega_f^2 \langle v^s, Mu^s \rangle,$$
$$\langle v^1, Uu^s \rangle = \omega_f^2 \langle v^1, Mu^s \rangle,$$
$$\langle v^s, Uu^1 \rangle = \omega_f^2 \langle v^s, Mu^1 \rangle,$$

for $s = 2, 3, \ldots, n$.

Furthermore, if $A(\omega^2) \equiv U - \omega^2 M$ is considered as a function of ω^2 so that η becomes a function of ω^2, then the first and second derivatives of η with respect to ω^2 and evaluated at $\eta = \eta_1$, $\omega^2 = \omega_f^2$ are given by [100, 101]

$$\left.\frac{d\eta}{d\omega^2}\right|_{\substack{\eta=\eta_1 \\ \omega^2=\omega_f^2}} = \frac{\langle v^1, Mu^1 \rangle}{\langle v^1, Eu^1 \rangle} = 0 \tag{6.22}$$

$$\left.\frac{d^2\eta}{d(\omega^2)^2}\right|_{\substack{\eta=\eta_1 \\ \omega^2=\omega_f^2}} = 2 \sum_{s=2}^{n} \frac{\langle v^1, Mu^s \rangle \langle v^s, Mu^1 \rangle}{\eta_1 - \eta_s}. \tag{6.23}$$

The first derivative follows immediately from (6.19) and the derivation of the expression for the second derivative is left to the reader as an exercise (see Problem 1).

In order to investigate the extremum properties of the generalized Rayleigh quotient with respect to the flutter point η_1, it is convenient to write x and y in the form

$$x = u^1 + w \quad \text{and} \quad y = v^1 + z. \tag{6.24}$$

When this form is substituted into (6.17) one obtains, after using (6.11),

$$\langle v^1, Mw \rangle + \langle z, Mu^1 \rangle + \langle z, Mw \rangle = 0. \tag{6.25}$$

Also, the generalized Rayleigh quotient (6.18) can be written in the form

$$R(x, y) = R(u^1, v^1) + \frac{\langle z, (U - \omega_f^2 M - \eta_1 E) w\rangle}{\langle y, Ex\rangle} \tag{6.26}$$

with the use of (6.19)–(6.21), (6.24) and (6.25). If in addition w and z are expressed as

$$w = \alpha_i u^i, \qquad z = \beta_i v^i \tag{6.27}$$

then (6.26) becomes

$$R(x, y) = R(u^1, v^1) + h \tag{6.28}$$

where

$$h = \frac{1}{\langle y, Ex\rangle} \sum_{s=2}^{n} (\eta_s - \eta_1)\, \alpha_s \beta_s. \tag{6.29}$$

It is seen from (6.28) that the generalized Rayleigh quotient will have a local minimum (maximum) at u^1, v^1 if $h \geq 0$ ($h \leq 0$) for all x, y in a neighbourhood of u^1, v^1. Such extremum results will now be derived for the case of $n = 2$ with η_1 real and $\eta_1 \neq \eta_2$. Since η_1 is real, η_2 is necessarily real and so are the corresponding eigenvectors, and α_i and β_j in (6.27) will be taken as real numbers in what follows. One considers then only real x and y.

First, it is noted that (6.23) takes the form

$$\left.\frac{d^2\eta}{d\omega^2}\right|_{\substack{\eta=\eta_1\\ \omega^2=\omega_f^2}} = 2\frac{\langle v^1, Mu^2\rangle\langle v^2, Mu^1\rangle}{\eta_1 - \eta_2}. \tag{6.30}$$

Next, consider an asymptotic analysis in which w and z are proportional to a small parameter ϵ. One can then obtain

$$h = (\eta_2 - \eta_1)\, \alpha_2 \beta_2 \tag{6.31}$$

from (6.24), (6.28) and (6.29), and

$$\langle v^1, Mw\rangle + \langle z, Mu^1\rangle = 0 \tag{6.32}$$

from (6.25) as first order approximations. This last expression can be written as

$$\alpha_2 \langle v^1, Mu^2\rangle + \beta_2 \langle v^2, Mu^1\rangle = 0 \tag{6.33}$$

and multiplying by $\beta_2 \langle v^2, Mu^1\rangle$ leads to the inequality

$$\langle v^1, Mu^2\rangle\langle v^2, MU^1\rangle\, \alpha_2 \beta_2 \leq 0. \tag{6.34}$$

The following conclusion can then be deduced from expressions (6.28), (6.30), (6.31) and (6.34): *Subject to condition* (6.17) *with real x and*

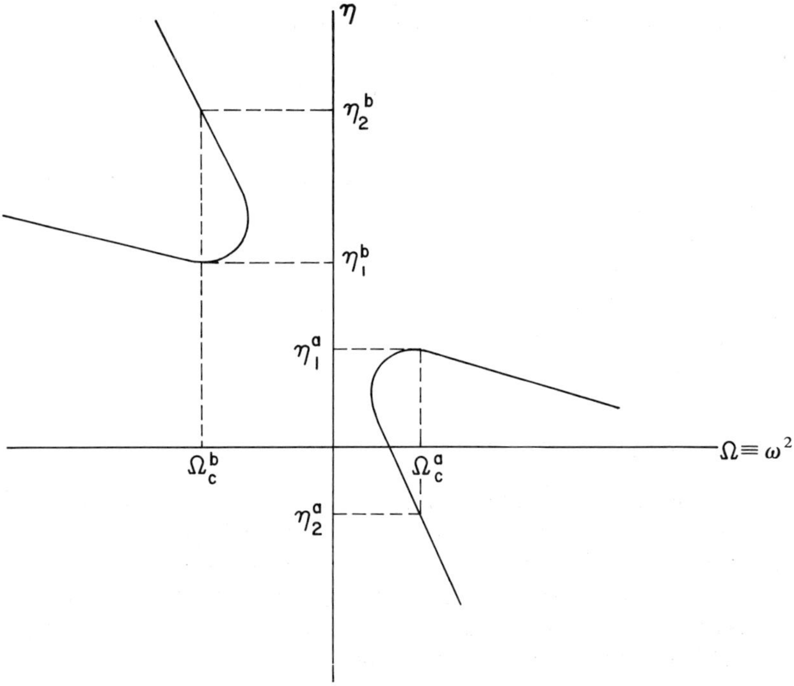

Figure 6.6

y, the generalized Rayleigh quotient (6.18) *has a local minimum* (*maximum*) *at* $x = u^1$, $y = u^1$ *if*

$$\left.\frac{d^2\eta}{d\omega^2}\right|_{\substack{\eta=\eta_1\\ \omega^2=\omega_f^2}} > 0 \; (<0). \tag{6.35}$$

This inequality can be related to the two values of η_1 at which flutter occurs. Let superscripts a and b designate the two flutter points. Then the characteristic equation

$$|-\omega^2 M + U - \eta E| = 0$$

has a double root in ω^2 at $\omega^2 = (\omega_f^2)_a$, $\eta = \eta_1^a$ and $\omega^2 = (\omega_f^2)_b$, $\eta = \eta_1^b$. A typical characteristic curve for the case $0 < \eta_1^a < \eta_1^b$ is depicted in Figure 6.6. It is a conic section and one can see that $d^2\eta/(d\omega^2)^2$ is negative at η_1^a and positive at η_1^b. The cases of eigenvalues with opposite sign and of two negative eigenvalues can be drawn similarly, and one obtains the following conclusions:

(i) if $0 < \eta_1^a < \eta_1^b$, then $R(x, y)$ has a local maximum at η_1^a and a local minimum at η_1^b;

(ii) if $\eta_1^a < 0 < \eta_1^b$, then $R(x, y)$ has a local minimum at η_1^a and a local maximum at η_1^b;

(iii) if $\eta_1^a < \eta_1^b < 0$, then $R(x, y)$ has a local maximum at η_1^a and a local minimum at η_1^b.

If one solves explicitly for η_1^a, η_1^b and the corresponding eigenvectors in terms of the elements of the matrices U, M and E, one can show [99] that the generalized Rayleigh quotient must lie in one of the shaded regions in Figure 6.7. That is, for cases (i) and (iii) $R(x, y)$ cannot lie between η_1^a and η_1^b, while for case (ii) it must lie between these two eigenvalues (see Problem 2).

The extremum properties derived above are in direct contrast to those for symmetric systems. If U and E are real symmetric matrices of order two, U is positive definite, E is nonsingular, and η_1^a, η_1^b are now defined as the eigenvalues of the pencil $U - \eta E$ with $\eta_1^a < \eta_1^b$, then the classical Rayleigh quotient

$$R(x) = \frac{\langle x, Ux \rangle}{\langle x, Ex \rangle}$$

can only take values in the *unshaded* regions of Figure 6.7.

As in the classical case, an approximate procedure can be used to compute eigenvalues and also to determine some bounds on the eigenvalues. If both critical values of η at flutter are positive, for example, then one can obtain a *lower bound* for the smaller critical value and an *upper bound* for the higher value by using, in (6.18), vectors x, y satisfying (6.17) as can be deduced from Figure 6.7 (i). A lower bound for η_1^a is especially useful since it assures that the system is stable for $\eta < \eta_1^a$. If one attempts to approach the critical value η_1^a by an iterative process, it will be approached from below according to present results, while the Rayleigh-Ritz procedure for symmetric systems approaches the smallest

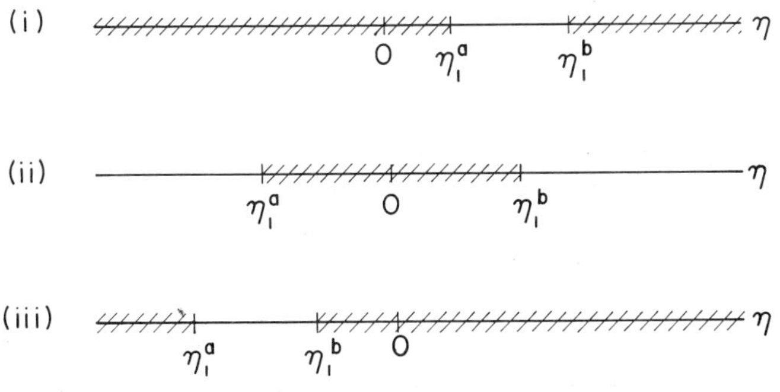

Figure 6.7

critical value from above and yields an upper bound rather than a lower bound.

6.3 Convexity of the flutter and divergence boundaries

The generalized Rayleigh quotient has an interesting interpretation for systems with multiple parameters, and the results of the preceding section can be used to establish certain convexity properties for the flutter boundary.

Considering again the multiple parameter system

$$(-\omega^2 M + U - \eta_k E^k)u = 0$$
$$v'(-\omega^2 M + U - \eta_k E^k) = 0 \tag{6.36}$$

one has

$$\langle v,(-\omega^2 M + U - \eta_k E^k)u\rangle = 0\,. \tag{6.37}$$

At a flutter point A, the eigenvectors x and y satisfy the flutter condition $\langle y, Mx\rangle = 0$ and (6.37) becomes

$$\langle y,(U - \eta_k E^k)x\rangle = 0 \tag{6.38}$$

which describes a plane tangent to the flutter boundary at A. In order to show this, note that a critical flutter point satisfies (6.36) and the flutter condition, and the critical values of η_k and ω^2 can, therefore, be expressed as functions of $(m-1)$ independent variables. Choosing η_j, $j \neq k$, as these $(m-1)$ variables, assuming a solution in the form $\eta_k(\eta_j)$ and $\omega^2(\eta_j)$, inserting this solution into (6.37) and differentiating the resulting equation with respect to η_j, one obtains

$$\left\langle \frac{\partial v}{\partial \eta_j},(-\omega^2 M + U - \eta_k E^k)u\right\rangle + \left\langle v,\left(\frac{\partial \omega^2}{\partial \eta_j}M - \frac{\partial \eta k}{\partial \eta_j}E^k - E^j\right)u\right\rangle$$
$$+\left\langle v,\left(-\omega^2 M + U - \eta_k E^k\right)\frac{\partial u}{\partial \eta_j}\right\rangle = 0. \tag{6.39}$$

The first and last terms in (6.39) are zero due to (6.36), and the evaluation at the flutter point A yields

$$\frac{\partial \eta_k}{\partial \eta_j} = -\frac{\langle y, E^j x\rangle}{\langle y, E^k x\rangle}\,. \tag{6.40}$$

But the plane (6.38) also has the same derivatives in arbitrary $\eta_k - \eta_j$ planes, and since the flutter point A satisfies (6.38) this plane is tangent to the flutter boundary at A. Futhermore, if n_k denotes the normal vector at A, it satisfies the the normality condition

$$\left(\eta_k^A - \eta_k\right) n_k = 0, \tag{6.41}$$

where η_k is any point on the tangent plane. Then, subtracting

$$\langle y, (U - \eta_k^A E^k)x \rangle = 0 \tag{6.42}$$

from (6.38) shows that the normal vector is given by

$$n_k = \langle y, E^k x \rangle. \tag{6.43}$$

Now consider a ray emerging from the origin of the η_k space and defined by

$$\eta_k = l_k \eta, \qquad \sum_{k=1}^{m} l_k^2 = 1. \tag{6.44}$$

Denoting $E = l_k E^k$, and assuming that the ray initially hits the flutter boundary at $\eta = \eta_1 > 0$ with corresponding eigenvectors u^1, v^1, one has

$$\eta_1 = \frac{\langle v^1, Uu^1 \rangle}{\langle v^1, Eu^1 \rangle} = R(u^1, v^1) \tag{6.45}$$

and

$$\eta_t = \frac{\langle y, Ux \rangle}{\langle y, Ex \rangle} = R(x, y) \tag{6.46}$$

where η_t represents the intersection of the ray with the tangent plane (6.38). In other words, the generalized Rayleigh quotient $R(x, y)$ gives the distance η_t from the origin to the intersection of the ray with the tangent plane, while its stationary value $R(u^1, v^1)$ is the distance η_1 to the flutter boundary. This is depicted in Figure 6.8, where the eigenvectors x, y correspond to point A and u^1, v^1 to point B.

According to the extremum properties of the generalized Rayleigh quotient established in Section 6.2, one has

$$\eta_t \le \eta_1$$

on the positive ray η, for a two-degree-of-freedom system. The following theorem can then be stated [98, 99]:

THEOREM 6.2. *The initial flutter boundary of a circulatory system with two degrees of freedom cannot have concavity toward the origin.*

This result can be of practical use in obtaining lower and upper bounds on the portion of the flutter boundary which yields initial flutter instability.

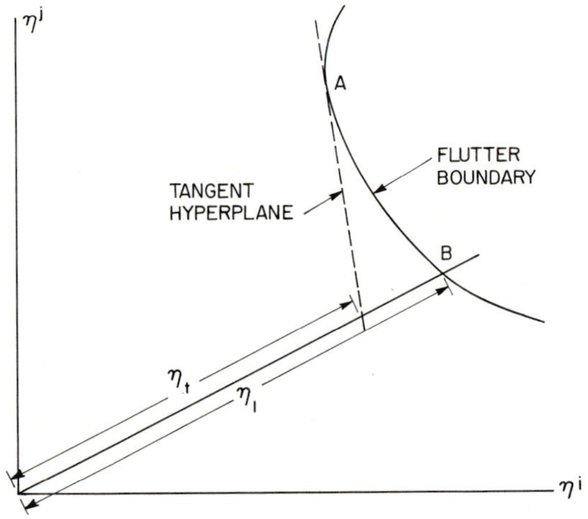

Figure 6.8

In addition, one can determine conditions under which the flutter boundary consists of planes. If the eigenvectors x and y at a flutter point are independent of the direction cosines of the ray to that point, then the normal vector with components given by (6.43) is constant for that portion of the flutter boundary and hence this boundary is linear. For a system with two degrees of freedom this condition is satisfied if in (6.36) $M = I$, U is a diagonal matrix, and the ratio of the off-diagonal terms for each matrix E^k is the same. This point will be elaborated in the following alternative analysis which will also lead to similar results for the divergence boundary.

An alternative method

The convexity properties of flutter and divergence boundaries can be explored by another interesting approach. The method will first be applied to the flutter boundary by considering a two-degree-of-freedom system with two independent parameters.

Let F be a critical flutter point on the characteristic surface (6.5) where $k = 1, 2$; the matrices M, U and E are all 2×2 matrices, and it is assumed without loss of generality that $M = I$ and U is diagonal. The coordinate axes are translated to F, Figure 6.9, by introducing

$$\eta_k = \eta_k^f + \bar{\eta}_k \qquad (6.47)$$
$$\omega^2 = \omega_f^2 + w$$

where $\bar{\eta}_k$ and w are incremental quantities. Substituting (6.47) into the characteristic equation (6.5) gives

$$V(w,\bar{\eta}_k) \equiv |-(\omega_f^2+w)I+A-\bar{\eta}_k E^k|=0 \tag{6.48}$$

where $A = U - \eta_k^f E^k$, and at point F, the flutter condition yields

$$(a_{11}-a_{22})^2+4a_{12}a_{21}=0 \tag{6.49}$$

where a_{ij} are the elements of A. (6.48) can now be expanded into a Taylor series to yield the asymptotic equation of the surface in the vicinity of F. The analysis will be facilitated, however, if one first introduces a linear non-singular transformation of the $\bar{\eta}_k$,

$$\bar{\eta}_k = t_{kj}\varphi_j\,, \qquad |T| \neq 0\,, \tag{6.50}$$

so that the resulting function

$$\Pi(w,\,\varphi_j) \equiv V(w,\,t_{kj}\varphi_j)$$

will have the following properties:

$$\Pi^1 \equiv \left.\frac{\partial \Pi}{\partial \varphi_1}\right|_F \neq 0 \qquad \Pi^2 \equiv \left.\frac{\partial \Pi}{\partial \varphi_2}\right|_F = 0. \tag{6.51}$$

Geometrically, this transformation rotates the $\bar{\eta}_k$ axes around w so that φ_1 becomes normal to the flutter boundary while φ_2 takes a tangential direction at F as shown in Figure 6.9(b).

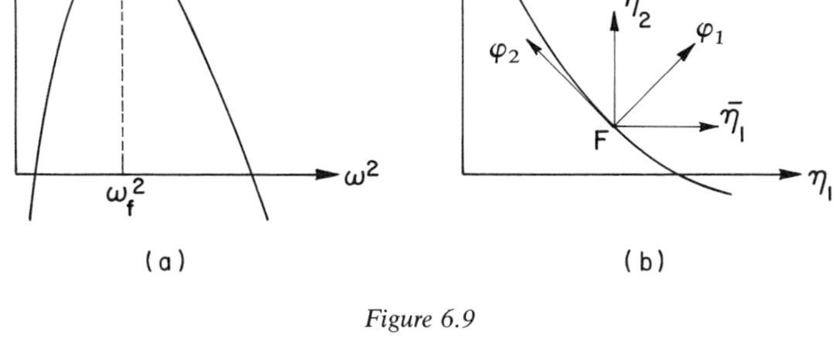

Figure 6.9

The characteristic equation now has the form

$$\Pi(w, \varphi_k) \equiv |-(\omega_f^2 + w)I + A - \varphi_k B^k| = 0$$

where (6.52)

$$B^k = t_{kl} E^l.$$

Expansion of $\Pi(w, \varphi_j)$ into a Taylor series around F yields

$$\Pi_0 + \Pi_1 w + \Pi^i \varphi_i + \tfrac{1}{2}\Pi_{11}(w)^2 + \Pi_1^i w\varphi_i + \tfrac{1}{2}\Pi^{ij}\varphi_i\varphi_j + \cdots = 0$$

where subscripts and superscripts on Π denote partial differentiation with respect to w and the φ_j respectively, all derivatives being evaluated at F. Since F is a critical flutter point, it follows that $\Pi_0 = 0$, $\Pi_1 = 0$ due to $\Delta = \partial\Delta/\partial\omega^2 = 0$, and $\Pi^2 = 0$ due to (6.51). The first order equation of the characteristic surface is, therefore, in the form

$$\Pi^1 \varphi_1 + \tfrac{1}{2}\Pi_{11}(w)^2 + \Pi_1^2 w\varphi_2 + \tfrac{1}{2}\Pi^{22}(\varphi_2)^2 = 0. \tag{6.53}$$

The coefficients of this equation can readily be obtained by differentiating the determinant (6.52) appropriately and evaluating at F as

$$\Pi_{11} = 2, \qquad \Pi_1^2 = b_{11}^2 + b_{22}^2, \qquad \Pi^{22} = 2(b_{11}^2 b_{22}^2 - b_{12}^2 b_{21}^2) \tag{6.54}$$

where b_{ij}^2 are the elements of B^2.

Before proceeding further, it is interesting to note that the characteristic surface (6.53) is analogous to *the equilibrium surface* in the vicinity of a *general critical point* associated with conservative systems (which is also called fold catastrophe), if the frequency w is treated as the behaviour variable and φ_j as control parameters [10, 12].

Returning to the analysis, one observes that the conditions $\Pi_1 = 0$ and $\Pi^2 = 0$ lead to the relations

$$2\omega_f^2 - (a_{11} + a_{22}) = 0 \tag{6.55}$$

and

$$b_{11}^2(\omega_f^2 - a_{22}) + b_{22}^2(\omega_f^2 - a_{11}) + a_{12}b_{11}^2 + a_{21}b_{12}^2 = 0 \tag{6.56}$$

respectively.

The first order equation of the flutter boundary can be obtained (just like the stability boundary on an equilibrium surface [10]) by differentiating (6.53) with respect to w, giving

$$w = -\frac{\Pi_1^2}{\Pi_{11}}\varphi_2, \tag{6.57}$$

and then substituting this value of w into (6.53). Thus, one obtains

$$\Pi^1\varphi_1 + \frac{1}{2\Pi_{11}}[\Pi_{11}\Pi^{22} - (\Pi_1^2)^2](\varphi_2)^2 = 0. \tag{6.58}$$

Using (6.54), (6.49), (6.55) and (6.56), the curvature of the flutter boundary at F can be expressed as

$$\left.\frac{\partial^2 \varphi_1}{\partial(\varphi_2)^2}\right| = \frac{2}{\Pi^1}\left(\frac{a_{12}b_{21}^2 - a_{21}b_{12}^2}{a_{11}-a_{22}}\right)^2. \tag{6.59}$$

Comparing this with the curvature

$$\left.\frac{\partial\varphi^2\varphi_1}{\partial w^2}\right|_F = -\frac{2}{\Pi^1} \tag{6.60}$$

obtained from (6.53), it follows that these curvatures always have opposite signs (Figure 6.9). This leads to the following theorem [98]:

THEOREM 6.3. *The flutter boundary is convex toward the region of real frequency* ω^2.

In particular, the corresponding portion of the stability boundary is convex toward the region of stability, which is equivalent to Theorem 6.2.

Furthermore, the *general critical point* here is *hyperbolic* [10, 12]; i.e., the flutter point F is always associated with anticlastic characteristic surfaces in terms of the general theory of stability originally developed for gradient systems.

A special case deserves further attention. Making use of (6.49), (6.52), (6.55) and (6.56), it can be shown that the curvature (6.59) vanishes if

$$e_{12}^1 e_{21}^2 = e_{12}^2 e_{21}^1 \tag{6.61}$$

where e_{ij}^k are the elements of the matrices E^k. The following theorem and corollary can, therefore, be stated:

THEOREM 6.4. *If* (6.61) *is satisfied, then the flutter boundary consists of straight lines.*

COROLLARY 6.2. *The flutter boundary consists of straight lines whenever one of the following conditions is satisfied:*

(*i*) *one of the matrices* E^k *is diagonal*;
(*ii*) $e_{ij}^k = -e_{ji}^k$ *for* $i \neq j$, $k = 1, 2$.

Divergence boundary

Let F now be a critical divergence point on the stability boundary satisfying the characteristic equation with $\omega^2 = 0$. Following the same

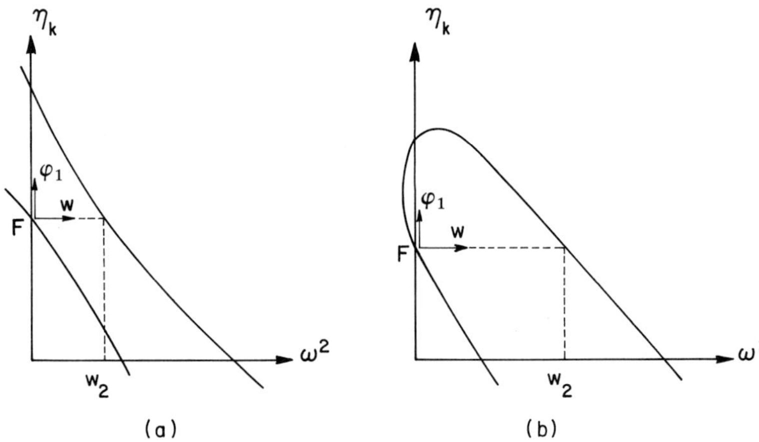

Figure 6.10

procedure as before, one obtains (6.48) with $\omega_f^2=0$ and the divergence condition

$$a_{11}a_{22}-a_{12}a_{21}=0. \tag{6.62}$$

The characteristic equation in transformed parameters (Fig. 6.10) is in the form

$$\Pi(w,\varphi_k)\equiv|-wI+A-\varphi_k B^k|=0 \tag{6.63}$$

When $\varphi_k=0$, (6.63) yields $w_1=0$ and $w_2>0$ since F is on the stability boundary. Thus, it will be assumed without loss of generality that

$$a_{11}=a_{12}=a_{21}=0, \qquad a_{22}>0. \tag{6.64}$$

The Taylor series expansion of (6.63) yields the first order equation

$$\Pi^1\varphi_1+\Pi_1 w+\tfrac{1}{2}\Pi^{22}(\varphi_2)^2=0 \tag{6.65}$$

where

$$\Pi^1=-b_{11}^1 a_{22}, \qquad \Pi_1=-a_{22}, \qquad \Pi^{22}=-2b_{12}^2 b_{21}^2, \tag{6.66}$$

and the fact that $\Pi^2=-b_{11}^2 a_{22}=0$, resulting in $b_{11}^2=0$, has been used.

From (6.65) and (6.66) one can immediately determine the slope of the characteristic surface at F as

$$\left.\frac{\partial\varphi_1}{\partial w}\right|_F=-\frac{1}{b_{11}^1} \tag{6.67}$$

and the curvature of the divergence boundary (which is now also the stability boundary) as

$$\left.\frac{\partial^2 \varphi_1}{\partial(\varphi_2)^2}\right|_F = -\frac{2b^2_{12}b^2_{21}}{b^1_{11}a_{22}}. \tag{6.68}$$

Assuming that the slope (6.67) is negative, which amounts to directing φ_1 from the region of stability to instability (Figure 6.10) one obtains $b^1_{11} > 0$.

Next, consider the behaviour of the system when the φ_2 direction is followed ($\varphi_1 = 0$). Using $b^2_{11} = 0$, the characteristic determinant (6.63) yields

$$(w)^2 - (a_{22} - b^2_{22}\varphi_2)w - b^2_{12}b^2_{21}(\varphi_2)^2 = 0 \tag{6.69}$$

with the discriminant

$$(a_{22} - b^2_{22}\varphi_2)^2 + 4b^2_{12}b^2_{21}(\varphi_2)^2 = 0 \tag{6.70}$$

which show that

$$b^2_{12}b^2_{21} \le 0 \tag{6.71}$$

if the system is capable of flutter, since $b^2_{12}b^2_{21} > 0$ implies that only divergence is possible. It then follows that

$$\frac{\partial^2 \varphi_1}{\partial(\varphi_2)^2} \ge 0, \tag{6.72}$$

and the following theorem can be stated:

THEOREM 6.5. *The initial divergence boundary (i.e., the divergence portion of the stability boundary) cannot have concavity toward the region of stability.*

Combining Theorems 6.4 and 6.5, the following result is formulated for the systems under consideration:

THEOREM 6.6. *The stability boundary of a two-degree-of-freedom circulatory system cannot have concavity toward the region of stability.*

6.4 Examples and discussion

The simply supported column shown in Figure 6.1(a) was given as Exercise 4 in Chapter 4. This is a pseudo-conservative system, and it can be shown by using a two-mode Galerkin analysis that $a_{12}a_{21} > 0$, so that

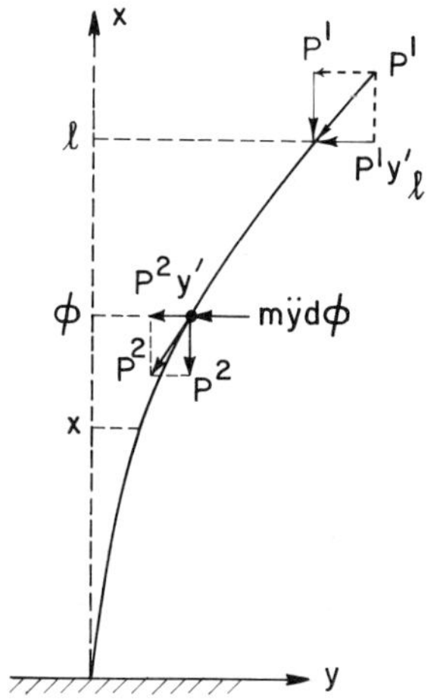

Figure 6.11

(6.49) rules out the possibility of flutter. The divergence boundary is shown (note that $\eta_k \equiv P^k$) in Figure 6.2 for comparison, and it is observed that the boundary is concave toward the origin according to the theory of Chapter 4.

The system shown in Figure 6.1(c) has been examined in [102]. P^1 is a follower load at the free end of the cantilever and P^2 represents a uniformly distributed tangential load. Both loads P^1 and P^2 lead to initial instability by flutter when applied individually. With the use of the first two natural modes of vibrations and Galerkin's procedure the continuous system reduces to a discrete one, and the application of the theory presented in this chapter leads to the divergence ($\omega^2 = 0$) and flutter boundaries (double root in ω^2) shown in Figure 6.4. Note that divergence instability can occur for certain combinations of P^1 and P^2.

The analysis of this system is carried out by applying the same method as in the other continuous examples. Thus, to obtain the differential equation of the cantilever shown in Figure 6.11, one considers the forces acting on an element $d\varphi$ and writes the bending moment at x,

$$\begin{aligned} sy''(x, t) = {} & P^1[y(l, t) - y(x, t)] - (l - x)P^1 y'(l, t) \\ & + \int_x^l P^2[y(\varphi, t) - y(x, t)]\, d\varphi - \int_x^l P^2 y'(\varphi, t)\, (\varphi - x)\, d\varphi \\ & - \int_x^l m\ddot{y}(\varphi, t)(\varphi - x)\, d\varphi. \end{aligned}$$

Differentiating this equation twice with respect to x yields the governing differential equation

$$sy'''' + [P^1 + P^2(l-x)]y'' + m\ddot{y} = 0 \tag{6.73}$$

where s denotes the constant flexural rigidity and m is the mass per unit length.

The boundary conditions are

$$y(0,t) = y'(0,t) = y''(l,t) = y'''(l,t) = 0.$$

A variational derivation of the differential equation (6.73) with P^1 and P^2 acting individually can be found in Leipholz's book [15].

Introducing now

$$y(x,t) = e^{i\omega t}z(x)$$

yields

$$sz'''' + [P^1 + P^2(l-x)]z'' - \omega^2 mz = 0 \tag{6.74}$$

with the boundary conditions

$$z(0) = z'(0) = z''(l) = z'''(l) = 0. \tag{6.75}$$

For a two-degree-of-freedom approximation, one then introduces

$$z = u_1 z_1(x) + u_2 z_2(x)$$

where $z_1(x)$ and $z_2(x)$ satisfy the boundary conditions (6.75), and can be chosen as the natural vibration modes [103]. (6.74) then yields the matrix equation (6.3) with the characteristic equation (6.5), and the application of the analysis described in this section results in the divergence and flutters boundaries shown in (Fig. 6.4).

The example shown in Figure 6.1(b) was analyzed in [104]. The distributed load P^1 represents the weight of the cantilevered column, while P^2 is the same as in the preceding example. When acting individually, P^1 is conservative and can cause divergence only, while P^2 leads to flutter. From a similar two-mode Galerkin approximation used in [104] one can obtain the boundaries depicted in Figure 6.3. Since flutter is possible under certain combinations of loading, the stability boundary (shaded) is convex toward the region of stability as proved in the preceding section (Theorem 6.6). Also note that the system is unstable for large positive values of P^1 and P^2, and the flutter and divergence boundaries contact each other tangentially at two points in the (P^1, P^2) plane. One of these points is in the first quadrant which is often of practical interest, while in the preceding example (Fig. 6.4) the contact points are outside this region.

The tangentiality of flutter and divergence boundaries is of course not an unexpected phenomenon if the following is observed: (i) the frequencies corresponding to the flutter and divergence regions are complex and real respectively, ruling out overlapping of these regions, and (ii) a gap between two regions would allow for a stable region extending to both positive and negative infinity (Figs. 6.3 and 6.4). This property can also be observed in the results of Herrmann and Bungay [105] who analyzed a two-degree-of-freedom model subjected to a partial follower force capable of producing both divergence and flutter instabilities and studied the connection between the two mechanisms.

Now consider the stability boundary in the first quadrant of the loading plane, which is often of greatest practical significance. The examples illustrate three typical forms of the stability boundary, depending on whether the loads P^1 and P^2, when acting individually in the positive direction, are such that (a) both can only lead to instability by divergence, (b) one leads to initial divergence instability while the other leads to flutter, or (c) both lead to flutter (Fig. 6.12). In each case lower and upper bound estimates (dashed lines in Fig. 6.12) can readily be obtained on the basis of the convexity theorems. For example, if both P^1 and P^2 lead to flutter (Case (c)), the straight line *AB* provides an upper bound while the tangent lines *AC* and *BC* give a lower bound. For Case (a) these upper and lower bounds are reversed. In (b) a lower bound *ACB* is obtained in the same manner as in (c), while the construction of the upper bound *ADB* involves the location *D* of contact between the flutter and divergence boundaries. The points *A* and *B* in Figure 6.12 represent the respective critical values of P^1 and P^2 when applied singly. The slopes of the boundaries are obtained from the stationarity property of the generalized Rayleigh quotient in the form of (6.40). It is noted that the

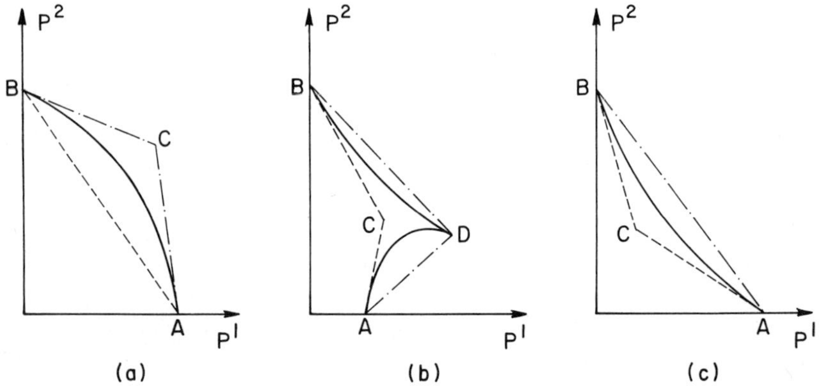

Figure 6.12

expressions for the slopes are similar in form for both divergence and flutter boundaries. This can be seen readily by considering the general equation

$$(-\omega^2 M + U - \eta_1 E^1 - \eta_2 E^2)u = 0$$

with the associated left eigenvector v. Pre-multiplying this equation by v', one observes that on the flutter boundary $\langle v, Mu\rangle = 0$ while on the divergence boundary $\omega^2 = 0$, resulting in

$$\frac{\mathrm{d}\eta_1}{\mathrm{d}\eta_2} = -\frac{\langle v, E^2 u\rangle}{\langle v, E^1 u\rangle}$$

in both cases. At the singular point D where the phenomenon of simultaneous flutter and divergence occurs, the slopes are the same and tangentiality of the boundaries follows. The point D can be located by solving (6.49) and (6.62) simultaneously.

6.5 A direct method of stability analysis of continuous system

One characteristic feature of continuous elastic systems subjected to circulatory forces is the nonself-adjointness of the differential operators associated with these forces. Here, self-adjointness is defined as in Chapter 1 except that instead of a matrix one now has a differential operator. Thus, a differential operator L is self-adjoint if

$$\langle \varphi_i, L\varphi_j\rangle = \langle \varphi_j, L\varphi_i\rangle$$

where φ_i and φ_j are certain functions. Otherwise the operator L is nonself-adjoint.

The eigenvalue analysis of circulatory systems associated with nonself-adjoint operators cannot be performed by the classical Rayleigh–Ritz procedure, and it is noted that Galerkin's procedure has been employed in the solutions of examples presented so far. Several authors [106–116] proposed modified variational techniques which are designed to overcome the difficulties arising from the nonself-adjoint operators. It seems that the concepts of *adjoint operators* and *adjoint systems* provide the most natural extension of Rayleigh's principle to nonconservative systems. In fact, the reader has already observed the significance of the generalized Rayleigh quotient and the concept of a left eigenvector which corresponds to considering an adjoint system in continuous systems. Here, a variational technique is developed which leads to divergence and flutter loads of elastic continuous systems directly. This technique is direct in the sense that it does not involve the frequencies of vibration, while Galerkin's procedure actually is concerned with a vibration analysis leading to the flutter load on the basis of an iterative procedure which ascertains the lowest value of the parameter at which two frequencies coalesce.

Consider an elastic system whose deflection modes $z(x)$ are governed by the differential equation [67]

$$-m(x)\omega^2 z(x) + Uz(x) + \eta Ez(x) = 0 \tag{6.76}$$

and the homogeneous boundary conditions

$$Bz(x) = 0 \tag{6.77}$$

where $\eta \geq 0$ is the control parameter (e.g., load), ω is the frequency, $m(x) > 0$ is the mass density, and U, E and B are linear differential operators in terms of x and are independent of ω^2 and t. It is also assumed that U and E are independent of η while B may be linear in η. The stability of the equilibrium state $z(x) \equiv 0$ is to be investigated.

Similar to considering the left eigenvectors (or the transposed adjoint system) in the matrix analysis, one can introduce the adjoint of (6.76) as

$$-m(x)\omega^2 w(x) + U'w(x) + \eta E'w(x) = 0 \tag{6.78}$$

with associated boundary conditions

$$Aw(x) = 0 \tag{6.79}$$

where the operators involved are such that

$$\int w(Uz + \eta Ez)\, \mathrm{d}x = \int z(U'w + \eta E'w)\, \mathrm{d}x \tag{6.80}$$

whenever (6.77) and (6.79) are satisfied. Normally $U = U'$; however, this assumption is not made here.

If (6.76) is multiplied by w and integrated over x, one obtains

$$-\omega^2 \int mwz\, \mathrm{d}x + \int wUz\, \mathrm{d}x + \eta \int wEz\, \mathrm{d}x = 0 \tag{6.81}$$

Integrating the middle term by parts or using Green's formula results in an equation of the form

$$-\omega^2 \int mwz\, \mathrm{d}x + \int (L_1 w)(L_2 z)\, \mathrm{d}x + \eta \left[(A_1 w)(B_1 z) + \int wEz\, \mathrm{d}x \right] = 0. \tag{6.82}$$

Here A_1 and B_1 are operators on the boundary of the system and can occur when B is linear in η. Solving for η yields

$$\eta = -\frac{-\omega^2 \int mwz \, dx + \int (L_1 w)(L_2 z) \, dx}{(A_1 w)(B_1 z) + \int wEz \, dx} \tag{6.83}$$

With the operators L_1 and L_2 in appropriate form, the expression in (6.83) is stationary with respect to kinematically admissible variations in the functions z and w (see the examples later).

Assume that the mode z is associated with initial instability of the system. If instability occurs by divergence, then $\omega^2 = 0$. If flutter takes place, one can show that [114]

$$\int mwz \, dx = 0 \tag{6.84}$$

which corresponds to the flutter condition (6.17). To show (6.84), differentiate (6.83) to get

$$\frac{d\eta}{d\omega^2} = \frac{-\int mwz \, dx}{(A_1 w)(B_1 z) + \int wEz \, dx} \tag{6.85}$$

for the slopes of the characteristic curves. Since at flutter $d\eta/d\omega^2$ vanishes, (6.84) follows.

Thus, the expression

$$\eta = -\frac{\int (L_1 w)(L_2 z) \, dx}{(A_1 w)(B_1 z) + \int wEz \, dx} \tag{6.86}$$

is stationary with respect to kinematically admissible variations in z and w at the divergence points and is stationary with respect to kinematically admissible variations in z and w satisfying (6.84) at the flutter points. It is seen that this *variational* principle is independent of the frequency ω^2 and may form the basis of a direct technique for determining approximate critical loads in elastic systems with circulatory parameters.

Application

Let z and w be approximated by

$$z(x) = u_k z_k(x), \qquad w(x) = v_k w_k(x) \tag{6.87}$$

where $k = 1, 2, \ldots, n$, and z_k and w_k are two sets of linearly independent functions satisfying the kinematic boundary conditions in (6.77) and (6.79) respectively.

Substituting (6.87) into (6.83) yields

$$\eta = -\frac{-\omega^2\langle v, Mu\rangle + \langle v, Uu\rangle}{\langle v, Eu\rangle} \tag{6.88}$$

where M, U and E now denote $n \times n$ matrices given by

$$m_{jk} = \int m z_j w_k \; \mathrm{d}x,$$

$$U \equiv [a_{jk}] = \left[\int (L_2 z_j)(L_1 w_k)\, \mathrm{d}x\right],$$

$$e_{jk} = (B_1 z_j)\,(A_1 w_k) + \int (E z_j)\; w_k \; \mathrm{d}x. \tag{6.89}$$

It is emphasized once more that although the same notation is being used both for differential and matrix operators the context and definitions make the notation clear.

The flutter condition (6.84) takes the form

$$\langle v, Mu\rangle = 0. \tag{6.90}$$

It must be noted that u and v are yet to be adjusted on the basis of the variational principle established earlier. Thus, in order to derive approximate divergence loads from (6.88) with $\omega^2 = 0$, one can define the function

$$\begin{aligned} V &= \langle v, (U + \eta E)u\rangle \\ &\equiv (a_{jk} + \eta e_{jk}) u_j v_k \end{aligned} \tag{6.91}$$

and set

$$\frac{\partial V}{\partial u_j} = 0, \qquad j = 1, 2, \ldots, n \tag{6.92}$$

to obtain n linear equations in v_k. For a nontrivial solution,

$$|a_{jk} + \eta e_{jk}| = 0 \tag{6.93}$$

which is a polynomial of order n in the parameter η, and its solution yields approximate values of η for divergence instability.

Flutter instability can be investigated either by eliminating u_j and v_k through the application of the variational principle for (6.83) and then applying an alternate flutter condition to (6.90), or by a variational technique which retains all variables u_j and v_k and makes use of (6.90). In the first procedure, define

$$V = \langle v, (U + \eta E - \omega^2 M)u \rangle = 0$$
$$\equiv (a_{jk} + \eta e_{jk} - \omega^2 m_{jk}) u_j v_k = 0. \tag{6.94}$$

Due to the variational principle associated with (6.83), one considers

$$\frac{\partial V}{\partial u_j} = 0, \qquad j = 1, 2, \ldots, n \tag{6.95}$$

which lead to the characteristic equation

$$\Delta(\omega^2, \eta) \equiv |U + \eta E - \omega^2 M| = 0 \tag{6.96}$$

The solution of (6.96) and

$$\frac{\partial \Delta}{\partial \omega^2} = 0 \tag{6.97}$$

yields the flutter loads. For $n = 2$ or 3 one can solve (6.97) for ω^2 analytically and substitute into (6.96) to get a single equation in η. For high values of n, however, a numerical technique is required. One way of combining (6.96) and (6.97) by eliminating ω^2 can be described in an example in general terms as follows:

Suppose

$$\Delta(\lambda) = b_0\lambda^4 + b_1\lambda^3 + b_2\lambda^2 + b_3\lambda + b_4 \tag{6.98}$$

where $\lambda \equiv \omega^2$ is used for simplicity. Then

$$\Delta'(\lambda) \equiv \frac{\partial \Delta}{\partial \lambda} = 4b_0\lambda^3 + 3b_1\lambda^2 + 2b_2\lambda + b_3 \tag{6.99}$$

Since both

$$\Delta(\lambda) = 0 \quad \text{and} \quad \Delta'(\lambda) = 0 \tag{6.100}$$

one can write

$$\lambda^3\Delta(\lambda) = \lambda^2\Delta(\lambda) = \lambda\Delta(\lambda) = 9!,$$

and

$$\lambda^4\Delta'(\lambda) = \lambda^3\Delta'(\lambda) = \lambda^2\Delta'(\lambda) = \lambda\Delta(\lambda) = 0$$

which can be expressed in the matrix form

$$Bx = 0 \tag{6.101}$$

where

$$B=\begin{bmatrix} b_0 & b_1 & b_2 & b_3 & b_4 & 0 & 0 \\ 0 & b_0 & b_1 & b_2 & b_3 & b_4 & 0 \\ 0 & 0 & b_0 & b_1 & b_2 & b_3 & b_4 \\ 4b_0 & 3b_1 & 2b_2 & b_3 & 0 & 0 & 0 \\ 0 & 4b_0 & 3b_1 & 2b_2 & b_3 & 0 & 0 \\ 0 & 0 & 4b_0 & 3b_1 & 2b_2 & b_3 & 0 \\ 0 & 0 & 0 & 4b_0 & 3b_1 & 2b_2 & b_3 \end{bmatrix}, \qquad x=\begin{bmatrix} \lambda^7 \\ \lambda^6 \\ \lambda^5 \\ \lambda^4 \\ \lambda^3 \\ \lambda^2 \\ \lambda \end{bmatrix}$$

For a nontrivial solution of (6.101), B must be singular, and the flutter boundary is thus expressed by the single determinantal equation

$$|B|=0 \qquad (6.102)$$

which yields a polynomial in η.

The variational principle associated with (6.84) and (6.86) leads to an alternative procedure.

From the flutter condition (6.90), and equations (6.91) and (6.94), it is seen that the latter function V can be viewed as being constructed from the former by adding (6.90) with the *Lagrange multiplier* $-\omega^2$. The conditions for stationarity are then

$$\frac{\partial V}{\partial u_j}=0, \qquad \frac{\partial V}{\partial v_k}=0, \qquad \frac{\partial V}{\partial \omega^2}=0; \qquad j,k=1,2,\ldots,n \qquad (6.103)$$

which together with (6.94) can be solved for the unknowns u_j, v_k, η and ω^2. Various forms of solution methods of (6.103) can then be devised [67]. It is observed that treating $-\omega^2$ as a Lagrange multiplier provides for a symmetric treatment of the problem.

Example 1

Consider a uniform cantilevered column subjected to a partial follower load at its free end (Fig. 6.13). The governing equation can be derived from (6.73) as

$$sy'''' + \eta y'' + m\ddot{y} = 0$$

or from (6.74) as

$$sz'''' + \eta z'' - \omega^2 mz = 0 \qquad (6.104)$$

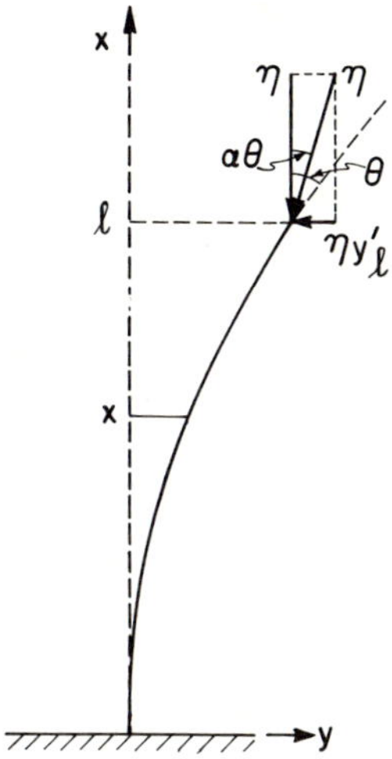

Figure 6.13

with the boundary conditions

$$z(0) = z'(0) = z''(l) = 0$$
$$sz'''(l) + (1-\alpha)\eta z'(l) = 0. \tag{6.105}$$

For $\alpha = 1$ the load acts tangentially, which yields the so-called Beck's problem. For $\alpha = 0$ it acts in a vertical direction (conservative). This system was studied in [117].

The adjoint system has boundary conditions

$$w(0) = w'(0) = 0,$$
$$sw''(l) + \alpha\eta w(l) = 0,$$
$$sw'''(l) + \eta w'(l) = 0, \tag{6.106}$$

and the same equation as for $z(x)$.

(6,83) takes the form

$$\eta = -\frac{-\omega^2 m \int_0^l wz \,\mathrm{d}x + s\int w''z'' \,\mathrm{d}x}{(\alpha - 1)w(l)z'(l) + \int_0^l wz'' \,\mathrm{d}x} \tag{6.107}$$

which is stationary with respect to variations in z and w satisfying the boundary conditions at $x = 0$.

The functions $z_j(x)$ and $w_k(x)$ in (6.87) are chosen such that the conditions

$$\begin{aligned} z_j(0) &= z_j'(0) = 0 \\ w_k(0) &= w_k'(0) = 0 \end{aligned} \tag{6.108}$$

are satisfied, and this is one of the advantages of the method proposed in this section. *In other words, just like in the Rayleigh–Ritz procedure, the deflection functions have to satisfy the kinematic (or geometric) boundary conditions only*, while in a Galerkin type procedure deflection functions must satisfy all boundary conditions.

For instance, let

$$z_j(x) = w_j(x) = \left(\frac{x}{l}\right)^{j+1}. \tag{6.109}$$

The quantities defined in (6.89) become

$$\begin{aligned} m_{jk} &= \frac{ml}{j+k+3} \\ a_{jk} &= \frac{sjk(j+1)(k+1)}{(j+k-1)l^3} \\ e_{jk} &= \frac{(j+1)[(j+k+1)\alpha-(k\times 1)]}{(j+k+1)l} \end{aligned} \tag{6.110}$$

An approximate value for the critical divergence load can then be obtained from (6.93), while (6.102) or (6.103) leads to approximate flutter loads. If $n=2$, for example, one finds

$$\frac{l^2}{s}\eta = \frac{4[13-15\alpha-(124-360\alpha+225\alpha^2)^{\frac{1}{2}}]}{3-2\alpha} \tag{6.111}$$

for divergence and

$$\frac{l^2}{s}\eta = \frac{8{\cdot}16-8{\cdot}27\alpha-(-8{\cdot}52+45{\cdot}1\alpha-31{\cdot}0\alpha^2)^{\frac{1}{2}}}{0{\cdot}414-0{\cdot}994\alpha+0{\cdot}548\alpha^2} \tag{6.112}$$

for flutter. (6.111) gives the values 2·48 and 10 at $\alpha=0$ and $\alpha=0{\cdot}5$ respectively, as compared to the exact critical loads $\pi^2/4$ and π^2. At $\alpha=0{\cdot}5$ and $\alpha=1$, (6.112) yields 28·0 and 77·2 while the exact values are 16·05 and 20·05, indicating that more modes are required in this case. With $n=4$, in fact, numerical computations furnish the values 16·13 and 20·40 respectively.

If functions satisfying certain orthogonality properties are chosen for z_j and w_k, calculations can further be simplified [67].

Example 2

Consider ncxt a flat elastic panel of infinite span subjected to a supersonic airflow over one side and a uniform midplane compressive force. In dimensionless form the governing differential equation is derived in [118] as

$$-\omega^2 z(x) + z''''(x) + \eta_1 z''(x) + \eta_2 z'(x) = 0$$

with simply supported boundary conditions

$$z(0)=0, \qquad z''(0)=0, \qquad z(1)=0, \qquad z''(1)=0.$$

Here η_1 is the loading parameter and η_2 is the dynamic pressure parameter. The adjoint system is the same except that η_2 is replaced by $-\eta_2$ in the differential equation. For this problem, one has

$$\eta_1 = -\frac{\int_0^1 w''z''\,dx + \eta_2 \int_0^1 wz'\,dx}{\int_0^1 wz''\,dx}$$

which is stationary with respect to variations in z and w satisfying $z = w = 0$ at $x = 0$ and $x = 1$.

Let

$$z_j(x) = w_j(x) = \sin j\pi x$$

so that

$$m_{jk} = 0{\cdot}5\delta_{jk}$$

$$a_{jk} = 0{\cdot}5(j\pi)^4\delta_{jk} + j\eta_2\gamma_{jk}$$

$$e_{jk} = -0{\cdot}5(j\pi)^2\delta_{jk}$$

where

$$\gamma_{jk} = \begin{cases} 0 & \text{if } (k-j) \text{ is even} \\ \dfrac{k}{(k^2-j^2)} & \text{if } (k-j) \text{ is odd.} \end{cases}$$

If $n = 2$, (6.93) yields the divergence boundary

$$\eta_1 = 2{\cdot}5\pi^2 \pm 0{\cdot}5\left[9\pi^4 - \frac{64(\eta_2)^2}{9\pi^4}\right]^{\frac{1}{2}}$$

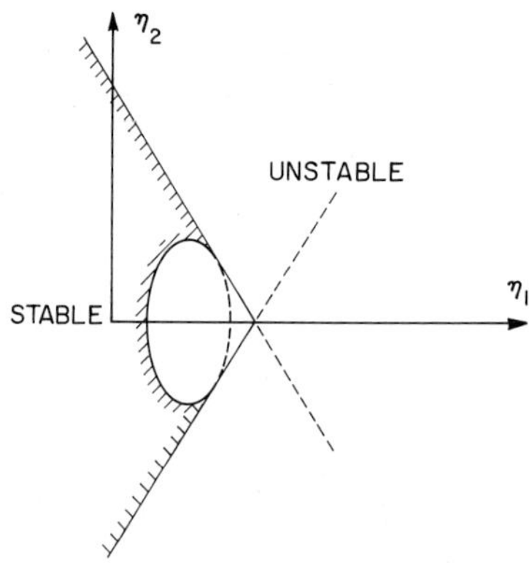

Figure 6.14

while (6.103) leads to

$$\eta_1 = \frac{\left[45\pi^4 - 8\eta_2\delta - \left(\frac{8\eta_2}{\delta}\right)\right]}{9\pi^2}$$

where $\delta = z_2/z_1$, and the stationarity condition $\partial\eta_1/\partial\delta = 0$ yields the flutter boundary

$$\eta_1 = 5\pi^2 \pm \left(\frac{16\eta_2}{9\pi^2}\right).$$

The divergence and flutter boundaries are shown in Figure 6.14 where the stability boundary is shaded. Note that convexity theorems established in the preceding section apply.

Problems

1. Derive the expression (6.23) (see [100]).

2. Solving explicitly for η_1^a, η_1^b and the corresponding eigenvectors in terms of the elements of the matrices U, M and E, show that the local extremum properties of the generalized Rayleigh quotient derived for a two-degree-of-freedom system in Section 6.2 hold globally and verify the results shown in (Fig. 6.7) (see [97]).

3. Analyze the problem shown in Figure 6.1(b) by the Galerkin method and show that the divergence and flutter boundaries are as shown in Figure 6.3.

4. Apply the direct method of stability analysis described in Section 6.5 to the same system as in Problem 3 and compare the two methods and the results.

5. Analyze the system shown in Figure 6.1(c) by the direct method of Section 6.5.

6.6 Effect of velocity-dependent forces

The effect of damping on the stability behaviour of circulatory systems attracted attention following the pioneering work of Ziegler [119] which led to the surprising result that internal damping can have a destabilizing effect on circulatory systems. Nemat-Nasser and Herrmann [120] showed that all sufficiently small velocity-dependent forces (and not only internal damping) may have a destabilizing effect on discrete systems. Prasad and Herrmann [121] proved certain theorems regarding the behaviour of damped circulatory systems. Jong showed [122] that the flutter boundary of an undamped two-degree-of-freedom system is the envelope of the flutter boundary of the corresponding system with slight internal damping. Other studies include [123–129].

Adding velocity-dependent forces to the circulatory system (6.2), one obtains

$$M\ddot{q}+(D+G)\dot{q}+Kq=0 \tag{6.113}$$

where $D\geq 0$ and $G=-G'$. It will be assumed that the damping represented by the positive semi-definite matrix D is pervasive as discussed in the previous chapters. Introducing $q=ue^{\lambda t}$ into (6.113) yields

$$[\lambda^2 M+\lambda(D+G)+K]u=0. \tag{6.114}$$

Pre-multiplying by $\bar{u}'$, one obtains

$$\lambda^2+(d+ig)+k+is=0 \tag{6.115}$$

with

$$\langle\bar{u},\, Mu\rangle=1, \qquad d=\langle\bar{u},\, Du\rangle>0,$$
$$ig=\langle\bar{u},\, Gu\rangle, \qquad \langle\bar{u},\, Ku\rangle=k+is,$$

where k is associated with the symmetric part of K, and s arises from the skew-symmetric part of K.

Introducing now $\lambda=\gamma+i\omega$ results in

$$\gamma^2-\omega^2+\gamma d-\omega g+k=0 \tag{6.116}$$

$$2\gamma\omega+\gamma g+\omega d+s=0 \tag{6.117}$$

which also follows from (5.136) with appropriate interpretation of the terms involved. The close connection between the systems under consideration and gyroscopic systems with internal and external damping is obvious. Hence, the stability criteria given in Section 5.10 can be adopted here as well, and it follows that the system (6.114) is asymptotically stable if and only if

$$d>0 \quad \text{and} \quad kd^2+sdg-s^2>0 \tag{6.118}$$

for all eigenvectors normalized with respect to M. Similarly, the equivalent criteria (5.139) and (5.140) are also applicable. The criticality condition (for $d>0$) is obtained when

$$kd^2+sdg-s^2=0 \tag{6.119}$$

for some mode(s). If (6.118) holds for all modes except the one which leads to (6.119), the critical point is *primary*. It was shown at the end of Section 5.10 that the above criterion has to be modified in the case of $D\equiv 0$, and that the system (6.113) is then generally unstable.

To show the destabilizing effect of slight damping, suppose the velocity-dependent terms in (6.113) are proportional to a small parameter ϵ so that one has

$$(\lambda^2 M+\lambda\epsilon(D+G)+K)u=0 \tag{6.120}$$

The undamped system ($\epsilon=0$) was shown to be stable when (6.7) holds, and it was observed that in the stability region eigenvectors are real so that (6.8) holds. It must be noted immediately, however, that the eigenvectors are not in general real.

If one considers the stable region of the undamped system with real eigenvectors u_r, the slightly damped system ($\epsilon\neq 0$) may be assumed to have vectors of the form

$$u=u_r+i\epsilon v \tag{6.121}$$

where v is real, and as ϵ tends to zero u tends to u_r. Since

$$\begin{aligned}
\langle \bar{u}, \epsilon Du\rangle &= \epsilon[\langle u_r, Du_r\rangle+\epsilon^2 \langle v, Dv\rangle]\equiv\epsilon(d_1+\epsilon^2 d_2)\\
\langle \bar{u}, \epsilon Gu\rangle &= \epsilon[2\epsilon i\langle u_r, Gv\rangle]\equiv\epsilon^2 ig\\
\langle \bar{u}, Ku\rangle &= \langle u_r, K_s u_r\rangle+\epsilon^2\langle v, K_s v\rangle+2\epsilon i\langle u_r, K_k v\rangle\\
&\equiv k_1+\epsilon^2 k_2+i\epsilon s
\end{aligned} \tag{6.122}$$

where K_s and K_k are the symmetric and skew-symmetric parts of K respectively, the stability criterion (6.118) takes the form [53]

$$d_1>0, \qquad k_1 d_1^2-s^2>0 \tag{6.123}$$

as a first order approximation ($\epsilon \to 0$). Here d_1 and k_1 are in terms of u_r as in the case of the undamped system which is stable if and only if $k_1 > 0$ and $s = 0$. It then follows from (6.123) that the introduction of small velocity-dependent forces has a destabilizing effect as long as $s \neq 0$. It is immediately noted, however, that this is not always the case, i.e., velocity-dependent forces do not always have a destabilizing effect on the system [123, 124, 127, 128]. Consider, for example, the case in which the velocity-dependent forces are associated with the positive definite matrix M so that the system (6.120) takes the form

$$(\lambda^2 M + \lambda \epsilon M + K)u = 0. \tag{6.124}$$

where ϵ is a positive real number.

It is easy to show that if this system is stable for $\epsilon = 0$, it is also stable for $\epsilon \neq 0$. In fact, if the undamped system is stable, then the matrix $M^{-1}K$ is *symmetrizable* and there exists a positive definite matrix S_1 such that $S_1 M^{-1} K$ is symmetric and positive definite (Theorem 6.1). Premultiplying (6.124) by M^{-1} and S_1 one obtains

$$(\lambda^2 S_1 + \lambda \epsilon S_1 + S_2)u = 0 \tag{6.125}$$

where $S_2 = S_1 M^{-1} K$. It then follows from Section 3.5 that the system (6.125) is asymptotically stable.

It is thus observed that the flutter boundary of a circulatory system is normally altered with the introduction of velocity-dependent forces. The divergence boundary, on the other hand, is defined by $\lambda = 0$ and it remains unaffected by such forces, but it may no longer be associated with the stability boundary since the destabilizing effect indicated by (6.123) applies also to divergence instability, flutter taking place before divergence. Such flutter points, however, are now included in the initial flutter boundary.

References

[1] Gantmacher, F. R., The Theory of Matrices, Vols. 1 and 2. Chelsea, New York (1959).

[2] Bellman, R., Introduction to Matrix Analysis. McGraw-Hill, New York, 2nd Edition (1970).

[3] Lancaster, P., Theory of Matrices. Academic Press, New York and London (1969).

[4] Taussky, O. and Zassenhaus, H., On the similarity transformation between a matrix and its transpose. *Pac. J. Math.* 9, 893–896 (1969).

[5] Taussky, O., Positive definite matrices and their role in the study of characteristic roots of general matrices. *Advances in Mathematics* 2, 175–186 (1968).

[6] Ostrowski, A. M., On the convergence of the Rayleigh quotient iteration for the computation of characteristic roots and vectors. *Arch. Rat. Mech. Anal. 1*, 233–241 (Part I); *2*, 423–428 (Part II); *3*, 325–340 (Part III): *3*, 341–347 (Part IV); *3*, 472–481 (Part V); *4*, 153–165 (Part IV) (1957–1960).

[7] Ostrowski, A. M., A quantitative formulation of Sylvester's law of interia. *Proc. Nat. Acad. Sci. U.S. 45*, 740–744 (1959).

[8] Shield, R. T. and Green, A. E., On certain methods in the stability theory of continuous systems. *Arch. Rat. Mech. Anal. 12*, 354 (1963).

[9] Koiter, W. T., On the concept of stability of equilibrium for continuous bodies. *Proc. R. Neth. Acad. Sci., Ser. [B] 66*, 173–177 (1963).

[10] Huseyin, K., Nonlinear Theory of Elastic Stability. Noordhoff International Publishing (1975).

[11] Thom, R., Structural Stability and Morphogenesis. Translated by D. H. Fowler. Benjamin, Inc. (1975).

[12] Huseyin, K., The multiple-parameter stability theory and its relation to catastrophe theory. *Proc. of the Interdisciplinary Conference on 'Problem Analysis in Science and Engineering*'. Edited by F. Branin and K. Huseyin. Academic Press (1977).

[13] Ziegler, H., Principles of Structural Stability. Blaisdell Publ. Co. (1968).

[14] LaSalle, J. and Lefschetz, S., Stability by Lyapunov's Direct Method. Academic Press, New York (1961).

[15] Leipholz, H. H. E., Stability Theory. Academic Press, Inc., New York and London (1970).

[16] Willems, J. L., Stability Theory of Dynamical Systems. Nelson (1970).

[17] Taussky, O. and Todd, J., On stable matrices. *Colloques Internationaux du Centre National de la Recherche Scientifique, No. 165*, 75–88 (1966).

[18] Hahn W., Stability of Motion. Springer–Verlag (1967).

[19] Chen, C. F. and Shieh, L. S., A note on expanding $PA+A^TP=Q$. *IEEE Trans. Automatic Control, AC-3*, 122–123 (1968).

[20] Muller, P. W., Solution of the matrix equation $AX+XB=-B$ and $S^TX+XS=-Q^*$. *SIAM J. Appl. Math. 18(3)*, 682–687 (1970).

References

[21] BARNETT, S. and STOREY, C., Some applications of the Lyapunov matrix equation. *J. Inst. Math. Appl. 4*, 33–42 (1970).

[22] PACE, I. S. and BARNETT, S., Comparison of numerical methods for solving Lyapunov matrix equations, *Intl. J. Control 15*, 907–915 (1972).

[23] ROTHSCHILD, D. and JAMESON, A. Comparison of four numerical algorithms for solving the Lyapunov matrix equation. *SIAM J. Appl. Math. 11*, 181–198 (1970).

[24] BEAVERS, Jr., A. N. and DENMAN, E. D., A new solution method for the Lyapunov matrix equations. *SIAM J. Appl. Math. 293*, (1976).

[25] CHETAYEV, N. G., The Stability of Motion. Translated by M. Nadler. Pergamon Press (1961).

[26] KOITER, W. T., On the instability of equilibrium in the absence of a minimum of the potential energy. *Proc. R. Neth. Acad. Sci., Ser. [B] 68*, 107–113 (1965).

[27] HAGEDORN, P., Die umkehrung der stabilitätssätze von Lagrange-Dirichlet ur. Routh. *Arch. Rat. Mech. Anal. 42, 4*, 281–316 (1971).

[28] PANOVKO, Y. G. and Gubanova, I. I., Stability and Oscillations of Elastic Systems. Translated by C. Vilarrick. $\frac{c}{b}$ Consultants Bureau (1965).

[29] HUSEYIN, K. and ROORDA, J., The loading-frequency relationship in multiple eigenvalue problems. *J. Appl. Mech. 38*, 1007–1011 (1971).

[30] PAPKOVICH, P. H., Works on the Structural Mechanics of Ships. Vol. 4 (1963).

[31] BUCKENS, F., Über Eigenwertscharen. *Öst. Ing. Arch. 12*, 83–93 (1958).

[32] HUSEYIN, K., The convexity of the stability boundary of symmetric structural systems. *Acta. Mech. 8*, 205–211 (1969).

[33] MASUR, E. F., Discussion on the paper 'The loading-frequency relationship in multiple eigenvalue problems'. *J. Appl. Mech. 39*, 636 (1972).

[34] HUSEYIN, K. and ROORDA, J., Closure on the discussion of E. F. Masur of 'The loading-frequency relationship in multiple eigenvalue problems'. *J. Appl. Mech.* 39, 639 (1972).

[35] DRUCKER, D. C., A more fundamental approach to plastic stress-strain relations. *Proc. First U.S. Nat. Cong. Appl. Mech., ASME*, 487–491 (1951).

[36] MASUR, E. F., Buclking, post-buckling and limit analysis of completely symmetric elastic structures. *Int. J. Solids Struct. 6*, 587–604 (1970).

[37] LANCASTER, P., Lambda-Matrices and Vibrating Systems. Pergamon Press (1966).

[38] MORAN, T. J., A simple alternative to the Routh-Hurwitz criterion for symmetric systems. *J. Appl. Mech. 37*, 1168–1170 (1970).

[39] WALKER, J. A. and SCHMITENDORF, W. E., A simple test for asymptotic stability in partially dissipative symmetric systems. *J. Appl. Mech. 40*, 1120–1121 (1973).

[40] LANCASTER, P., Bounds for latent roots in damped vibration problems. *SIAM Rev. 6(2)*, 121–125 (1964).

[41] MASUR, E. F., Bounds and error estimates in structural eigenvalue problems. *J. Struct. Mech. 1(4)*, 417–438 (1973).

[42] PEFLÜGER, A., Stabilitätsprobleme der Elastostatik. Springer-Verlag (1964).

[43] LEIPHOLZ, H. H. E., On the sufficiency of the energy criterion for the stability of certain non-conservative systems of the follower load type. *J. Appl. Mech. 39*, 717–722 (1972).

[44] HUSEYIN, K. and LEIPHOLZ, H. H. E., Divergence instability of multiple-parameter circulatory systems. *Quart. Appl. Math. 31(2)*. 185–197 (1973).

[45] LEIPHOLZ, H. H. E. and HUSEYIN, K., On the stability of one-dimensional continuous systems with polygenic forces. *Meccanica 6*, 253–257 (1971).

[46] HERRMANN, G., Dynamics and stability of mechanical systems with follower forces. NASA CR–1782 (1971).

[47] SUNDARARAJAN, C., On the stability of divergence-type elastic systems. *Devel. Theoret. Appl. Mech. 7*, Proc. Seventh Southeastern Conf. Theoret. Appl. Mech., Washington, D. C., 475–481 (1974).

[48] KORNECKI, A., On the character of instability of certain aeroelastic systems. *J. Appl. Mech. 40*, 616–617 (1973).

[49] HUSEYIN, K. and PLAUT, R. H., Application of the Rayleigh quotient to the eigenvalue problems of pseudo-conservative systems. *J. Sound. Vib. 33(2)*, 201–210 (1974).

[50] WALKER, J. A., Lyapunov analysis of the generalized Pflüger problem, *J. Appl. Mech.* 38, 935–938 (1972).

[51] MAGNUS, K., Der einfluss verschiedener kräftearten auf die stabilität linearer systeme. *ZAMP 21 (4)*, 523–534 (1970).

[52] BOTTEMA, O., On the stability of the equilibrium of a linear mechanical system. *ZAMP 6*, 97–103 (1955).

[53] SHIEH, R. C. and MASUR, E. F., Some general principles of dynamic instability of solid bodies. *ZAMP 19*, 927–941 (1968).

[54] HUSEYIN, K., Vibrations and stability of mechanical systems. *The Shock and Vibration Digest, Vol.* 8, *No. 4*, 56–66 (1975).

[55] THOMPSON, W. and TAIT, G., A treatise on natural philosophy. Vol. 1, Part 1, Cambridge University Press, Cambridge, England (1879).

[56] NIORDSON, F. E. (Ed.), Proceedings of IUTAM Symposium on Dynamics of Rotors, Lyngby, Denmark. Springer-Verlag (1975).

[57] BETH, H. J. H., On the stabilization of unstable equilibrium by means of gyroscopic forces. *Phil Mag. 49*, 447–450 (1925).

[58] SHIEH, R. C., Energy and variational principles for generalized (gyroscopic) conservative problems. *Intl. J. Nonlin. Mech.* 55, 495–509 (1971).

[59] HUSEYIN, K. and PLAUT, R. H., Transverse vibrations and stability of systems with gyroscopic forces. *J. Struct. Mech. 3(2)*, 163–177 (1974–75).

[60] HUSEYIN, K. and PLAUT, R. H., Divergence and flutter boundary of systems under combined conservative and gyroscopic forces. *Proc. IUTAM Symp. Dyn. Rotors*, Lyngby, Denmark. Springer-Verlag, 182–205 (1975).

[61] HUSEYIN, K., Standard forms of the eigenvalue problems associated with gyroscopic systems. *J. Sound Vib. 45(1)*, 29–37 (1976).

[62] MEIROVITCH, L., A new method of solution of the eigenvalue problem for gyroscopic systems. Seventh U.S. Natl. Congr. Appl. Mech., Boulder, Colo. (1974); *AIAAJ. 12(10)* 1337–1342 (1974).

[63] HAGEDORN, P., Über die instabilität konservativer systeme mit gyroskopischen kräften. *Arch. Rat. Mech. Anal. 58*, 1–9 (1975).

[64] TESCHNER, W., On the stability of gyroscopic systems. Doctoral Thesis, Darmstadt Technische Hochschule (1977).

[65] WALKER, J. A., On the stability of linear discrete dynamic systems. *J. Appl. Mech. 37*, 271–275 (1970).

[66] PLAUT, R. H., Alternative formulations for discrete gyroscopic eigenvalue problems. *AIAA J. 14(4)*, 431–435 (1976).

[67] PLAUT, R. H. and HUSEYIN, K., A direct method of stability analysis for elastic circulatory systems. *Developm. Mech., 13th Mid-West. Appl. Mech. Conf.*, 425–435 (1973).

[68] HOUSNER, G. W., Bending vibrations of a pipeline containing flowing fluid. *J. Appl. Mech., ASME 19*, 205–209 (1952).

[69] PLAUT, R. H. and HUSEYIN, K., Instability of fluid-conveying pipes under axial load. *J. Appl. Mech. 42*, 889–890 (1975).

[70] Paidoussis, M. P. and Issid, N. J., Dynamic stability of pipes conveying fluid. *J. Sound Vib. 33(3)*, 267–294 (1974).

[71] Weaver, D. S. and Unny, T. E., Dynamic stability of fluid-conveying pipes. *J. Appl. Mech. 40*, 48–52 (1973).

[72] Zajac, E. E., The Kelvin–Tait–Chetaev theorem and extensions. *J. Aeronaut. Sci. 11*, 46–49 (1964).

[73] Zajac, E. E., Comments on 'Stability of damped mechanical system', and a further extension. *AIAA J. 3*, 1749–1750 (1965).

[74] Pringle, Jr., R., Stability of damped mechanical systems. *AIAA J.* 2, 363–364 (1965).

[75] Pringle, Jr., R., On the stability of a body with connected moving parts. *AIAA J.* 4, 1395–1404 (1966).

[76] Roberson, R. E., Notes on the Thomson–Tait–Chetaev stability theorem. *J. Astronaut. Sci. 15*, 319–322 (1968).

[77] Connell, G. M., Asymptotic stability of second-order linear systems with semi-definite damping. *AIAA J. 7*, 1185–1187 (1969).

[78] Müller, P. C., Asymptotische stabilität von linearen mechanischen systemen mit positiv semidefiniter dämpfungs matrix. *ZAMM 52*, T197–T198 (1971).

[79] Kane, T. R. and Barba, P. M., Effects of energy dissipation on a spinning satellite. *AIAA J. 4(8)*, 1391–1394 (1966).

[80] Likins, P. W., Attitude stability criteria for dual-spin spacecraft. *J. Spacecraft Rockets 4 (12)*, 1638–1643 (1967).

[81] Likins, P. W., Stability analysis of mechanical systems with constraint damping. *AIAA J. 5(11)*, 2091–2094 (1967).

[82] Dimentberg, F. J., Flexural Vibrations of Rotating Shafts. Translated from Russian. Butterworths, London (1961).

[83] Bolotin, V. V., Nonconservative Problems of the Theory of Elastic Stability. Pergamon Press, New York (1963).

[84] Mingori, D. L., A stability theorem for mechanical systems with constraint damping. *J. Appl. Mech. 37*, 253–258 (1970).

[85] Crandall, S. H., The role of damping in vibration theory. *J. Sound Vib. 11(1)*, 3–18 (1970).

[86] Huseyin, K., The effect of damping on the flutter boundary of rotating systems. Int. Conf. Vibrations of Rotating Machinery. Held at Cambridge University 1976. *Prelim. proc. C*180/76, *I. Mech. E*, England, 133–138 (1976).

[87] Genin, J. and Maybee, J. S., External and material-damped three-dimensional rotor system. *Intl. J. Nonlin. Mech. 5*, 287–297 (1970).

[88] Ariaratnam, S. T., The vibration of unsymmetrical rotating shafts. *ASME Paper No.* 64–*WA*/*APM*–4 (1964).

[89] Pedersen, P. T., On self-excited whirl of rotors. *Ing. Arch. 42*, 267–284 (1973).

[90] Plaut, R. H., Stability of rotating shafts with axial load and damping. (to be published).

[91] Sethna, P. R. and Balachandra, M. B., On nonlinear gyroscopic systems. (to be published).

[92] Yazici, M., On the stability of dissipative gyroscopic systems. Thesis, University of Waterloo (1976).

[93] Gladwell, G. M. L. and Bishop, E. D., The receptances of uniform and nonuniform rotating shafts. *J. Mech. Eng. Scis. 11*, 78 (1959).

[94] Herrmann, G., Stability of equilibrium of elastic systems subjected to nonconservative forces. *Appl. Mech. Rev. 20*, 103–108 (1967).

[95] Sundararajan, C., The vibration and stability of elastic systems subjected to follower forces. *Shock Vib. Dig. 7(6)*, 89–105 (1975).

[96] Ashley, H., Aerolasticity. *Appl. Mech. Rev. 23*, 119–129 (1970).

[97] Nemat-Nasser, S., Thermoelastic stability under general loads. *Appl. Mech. Rev. 23*, 615–624 (1970).

[98] Huseyin, K. and Plaut, R. H., The elastic stability of two-parameter nonconservative systems. *J. Appl. Mech. 40(1)*, 175–180 (1973).

[99] Huseyin, K. and Plaut, R. H., Extremum properties of the generalized Rayleigh quotient associated with flutter instability. *Quart. Appl. Math. 32*, 189–201 (1974).

[100] Plaut, R. H. and Huseyin, K., Derivatives of eigenvalues and eigenvectors in non-self-adjoint systems. *AIAA J. 11(2)*, 250–251 (1973).

[101] Lancaster, P., On eigenvalues of matrices dependent on a parameter. *Num. Math. 6*, 377–387 (1964).

[102] Fisher, W., Die Knicklast des einseitig eingespannten Stabes als zweiparametriges, nicht-konservatives Eigenwertproblem. *ZAMP 20*, 554–547 (1969).

[103] Bishop, R. E. D. and Johnson, D. C., Vibration Analysis Tables. Cambridge Press, 42–47 (1956).

[104] McGill, D. J., Column instability under weight and follower loads. *J. Eng. Mech. Div., ASCE 97*, 629–635 (1971).

[105] Herrmann, G. and Bungay, R. W., On the stability of elastic systems subjected to nonconservative forces. *J. Appl. Mech. 31(3)*, 435–440 (1964).

[106] Roberts, P. H., Characteristic value problems posed by differential equations arising in hydrodynamics and hydromagnetics. *J. Math. Anal. and Appl. 1*, 195–219 (1960).

[107] Levinson, M., Application of the Galerkin and Ritz methods to non-conservative problems of elastic stability. *ZAMP 17*, 431–442 (1966).

[108] Prasad, S. N. and Herrmann, G., The usefulness of adjoint systems in solving nonconservative stability problems of elastic continua. *Intl. J. Solids Struct. 5*, 727–735 (1969).

[109] Prasad, S. N. and Herrmann, G., Adjoint variational methods in nonconservative stability problems. *Intl. J. Solids Struct. 8*, 29–40 (1972).

[110] Leipholz, H., Application of a generalized principle of Hamilton to nonconservative problems. *SM Report No. 38*, Solid Mechanics Division, University of Waterloo, Waterloo, Ontario (1970).

[111] Dubey, R., Variational method for nonconservative problems. *J. Appl. Mech. 37*, 133–136 (1970).

[112] Shieh, R. C., Variational method in the stability analysis of non-conservative problems. *ZAMP 21*, 88–100 (1970).

[113] Anderson, G. L., A comparison of approximate methods for solving non-conservative problems of elastic stability. *J. Sound Vib. 22*, 159–168 (1972).

[114] Plaut, R. H., Determining the nature of instability in nonconservative problems. *AIAA J. 10*, 967–968 (1972).

[115] Leipholz, H. H. E., On a stability theory of non-self-adjoint mechanical systems. *Acta Mech. 14*, 283–295 (1972).

[116] Nemat-Nasser, S., On circulatory force fields and their adjoints. *J. Appl. Mech. 35*, 829–831 (1968).

[117] Kordas, Z. and Zyczkowski, M., On the loss of stability of a rod under a super-tangential force. *Archiwum Mechaniki Stosawanej 1*, 7–31 (1963).

[118] Dugundji, J., Theoretical considerations of panel flutter at high supersonic Mach numbers. *AIAA J. 4*, 1257–1266 (1966).

References

[119] ZIEGLER, H., Die stabilitätskriterien der elastomechanik. *Ing. Arch. 20*, 49–56 (1952).

[120] NEMAT-NASSER, S. and HERRMANN, G., Some general considerations concerning the destabilizing effect in nonconservative systems. *ZAMP 17*, 305–313 (1966).

[121] PRASAD, S. N. and HERRMANN, G., Some Theorems on stability of discrete circulatory systems. *Acta Mech. 6*, 208–216 (1968).

[122] JONG, I. C., Some properties pertaining to the stability of circulatory systems. *Developm. Theor. Appl. Mech., Fifth Southeastern Conf. 5*, 207–234 (1971).

[123] PLAUT, R. H. and INFANTE, E. F., The effect of external damping on the stability of Beck's column. *Intl. J. Solids Struct. 6*, 491–496 (1970).

[124] PLAUT, R. H., A new destabilizing phenomenon in nonconservative systems. *ZAMM 51*, 319–321 (1971).

[125] URBANO, C., The effect of damping on the dynamic stability of nonconservative systems. *Meccanica 3*, 131–139 (1968).

[126] KAPOOR, R. N. and LEIPHOLZ, H. H. E., Flutter-divergence transition criteria in certain viscoelastic polygenic systems. *Dev. Mech., Proc. 13th Midwest Mech. Conf. 7*, 929–942 (1973).

[127] WALKER, J. A., A note on stabilizing damping configurations for linear nonconservative systems. *Intl. J. Solids Struct. 9*, 1543–1545 (1973).

[128] DONE, G. T. S., Damping configurations that have a stabilizing influence on non-conservative systems. *Int. J. Solids Struct. 9*, 203–215 (1973).

[129] BOLOTIN, V. V. and ZHINZHER, N. I., Effects of damping on stability of elastic systems subjected to non-conservative forces. *Int. J. Solids Struct. 5*, 965–989 (1969).

Index

Index